Sustainability: How the Cosmetics Industry is Greening Up

Sustainability: How the Cosmetics Industry is Greening up

Edited by

Amarjit Sahota
Organic Monitor, London, UK

This edition first published 2014

Registered office
John Wiley & Sons Ltd, The Atrium, Southern Gate, Chichester, West Sussex, PO19 8SQ, United Kingdom

For details of our global editorial offices, for customer services and for information about how to apply for permission to reuse the copyright material in this book please see our website at www.wiley.com.

Library of Congress Cataloging-in-Publication Data

Sustainability : how the cosmetics industry is greening up / edited by Mr Amarjit Sahota.
pages cm – (Advances in design and control)
Includes bibliographical references and index.
ISBN 978-1-119-94554-3 (cloth)
1. Cosmetics. 2. Green products. I. Sahota, Amarjit, editor of compilation.
TP983.S885 2014
646.7′2–dc23

2013020184

A catalogue record for this book is available from the British Library.

Typeset in 10.5/13pt Sabon by Aptara Inc., New Delhi, India

Contents

Preface

When I was approached about the prospects of a book on green issues in the cosmetics industry, I was apprehensive. I was apprehensive not about the prospects of the book, but about me: how could I find the time for such a book, considering I was running a business, frequently travelling to various parts of the world, whilst trying to juggle the demands of family life? After some thought, I reluctantly accepted the invitation for I considered it a service... it would be a way of sharing some of the knowledge we (the contributors) had accumulated to advance sustainability in the cosmetics industry.

It would be a service in terms of education and enlightenment. To many, sustainability and cosmetics are two terms that do not sit well together. Indeed, at a recent summit my company organised, one environmentalist questioned the term 'sustainable cosmetics'. He remarked, 'is it me, or does no one else see this as the elephant in the room... how can you call vanity products like cosmetics sustainable?' Whilst he was not entirely right, he had a point. Human consumption is far exceeding the rate of replenishment of the planet's resources. Should the Earth's resources not be re-directed to more meaningful products then cosmetics?

The role of cosmetics in society is often understated. Many products, like soaps, shampoo and toothpaste, are an essential part of basic hygiene. Other products, such as sun care creams and rash lotions, play an important role in skin protection and health. Even make-up products like mascara and foundation contribute to society by raising the confidence and improving the well-being of the wearers. Going back to ancient civilisations like that of the Egyptians, dyes were used as lipsticks and oils as fragrances. Thus, cosmetics have been and always will be an integral part of human society.

I took on this book project to describe how the cosmetics industry is tackling the sustainability challenge. The basic challenge faced by consumer product industries is – how do you continue to raise production levels whilst using increasingly finite resources? Consumption rates continue to rise whilst the Earth's resources become depleted. Over the last decade, we have seen prices of oil, minerals and agricultural commodities soar. Apart from the economic and ecological costs, there have been social repercussions, such as protests and riots over food inflation in developing countries. As economic output and human consumption continue to rise, the stress on the environment and society will inevitably increase.

What is peculiar about the cosmetics industry, however, is that it appears to be tackling this challenge relatively well. The industry is actually leading in many areas of sustainability; a number of surveys put cosmetic companies at the top of ethical corporation lists. In 2012, Natura Brasil was named the second most sustainable corporation in the world by Corporate Knights. Seven cosmetic companies were in the 2012 World's Most Ethical Companies List by Ethisphere Institute, up from four in 2011.

Cosmetic companies seem to be taking green issues seriously. Maybe it is because the industry has traditionally received negative media attention for animal-testing and unethical business practices? Maybe it is because the industry is heavily reliant on petroleum as feedstock for ingredients that companies are forced to look at green options? Or, perhaps, it is because cosmetic companies feel greater pressure to be sustainable because of the association of their products with vanity? Whatever the reason, green has become the new black in the cosmetics industry and there is much to learn from how the cosmetics industry is turning green.

The book has been written by industry experts, many of whom are from leading organisations involved in sustainability in the cosmetics industry. Over 14 chapters, the book describes the various ways cosmetic and ingredient firms are undertaking sustainability initiatives. Written by the industry, this book should be considered a practical guide for those looking to make a difference in terms of sustainability.

Amarjit Sahota
Organic Monitor, UK

Foreword

Sustainability plays a fundamental role in the cosmetics and personal care industry of today. Thinking and acting sustainably is an unquestioned priority for our future. However, this is not only an ethical commitment, it is also a business interest which must be embedded in the strategic planning of all companies, large and small.

You only need to look at the websites of many of our well-known company brands, available from our favourite stores, to see the important place that sustainability has in the industry's activities. Of course, we all have a special relationship with our cosmetic and personal care products – there are few consumer goods that we can name that we use on a daily basis, and that we change according to our changing needs and desires. Every single one of us uses a variety of cosmetic products every day of our lives. It is therefore important that cosmetic companies take the responsibility to think and act sustainably when it comes to sourcing, producing and using these millions of products valued throughout the world every day.

The social, economic and environmental pillars of sustainability all have a place at the heart of the life cycle of cosmetic and personal care products. The integration of these elements is not a marketing tool, it represents a genuine commitment to sustainability to be incorporated at many levels of the business model.

Sustainable thinking and practices have been embedded in company thinking for nearly three decades. Most companies publish annual sustainability reports, showing how companies are considering this element across many areas of work and throughout the product life cycle. Life cycle assessment is essential to take into account manufacturing, product formulation, packaging, distribution, the product consumption phase and the product post-use phase.

When it comes to company or association initiatives in the area of sustainable development, social and economic sustainability projects, as well as the more obvious environmental aspects, are given significant attention. This is true for projects within Europe, as well as throughout the world.

Regarding social sustainability, there are several important areas where companies and associations are active. We know that personal care products play a fundamental role in our well-being and self-esteem. Linking with this theme, there are initiatives that aim to help those who have lost their self-esteem or to increase awareness about its importance. Many companies also run different types of educational projects, for example training people in hairdressing in economically challenged communities, and enhancing understanding of the fundamental role of science as well as the important role of women in science. Medical projects exist to work with doctors providing facial surgery in remote communities. Raising awareness of dental hygiene is important for everyone – whether they are in communities where access to such care is not easy, or for the ageing population to ensure a good quality of life. Another aspect of older age that many companies encourage is that of enabling older people to continue to work.

The economic side to sustainability shows the fundamental importance of the cosmetics and personal care industry to our society's economic vibrancy. Not only does the industry contribute significantly to employment and to the economy, it also drives innovation and research using cutting edge advances. It is widely recognised that growth, competitiveness and employment are fundamental to the economic health of our society. So too is the economic contribution of small and medium sized enterprises (SMEs), which provide a power house of innovation and creativity that the cosmetic and personal care products industry greatly values.

Environmental sustainability is taken into account throughout the whole product life cycle, from sustainable sourcing of ingredients, to how the product is packaged and used. Companies are increasingly using special packaging materials, which can already be seen on our shelves. Manufacturing operations are also constantly being streamlined, with renewable energy sources increasingly being used. Reducing the carbon footprint is just one of many environmental elements taken into consideration in the industry. Studying environmental impacts from the supply chain to the use and final disposal of the product is essential in enabling us to confront the challenges of being environmentally sustainable.

As the voice of the cosmetics industry in Europe, Cosmetics Europe has recognised and understood the importance of the whole industry taking part in thinking and acting sustainably. Cosmetics Europe represents the interests of 18 multinational companies, over 4000 SMEs and 27 national associations. Together these companies and associations have produced 'Good Sustainability Practice Guidelines' and 'Ten Steps to Sustainability for SMEs', as well as providing a showcase of projects from all over the world illuminating how the cosmetics and personal care industry works across the three pillars of sustainability. At the level of the European Union, sustainable development is enshrined in the EU Treaty as one of the overarching principles of EU policy as well as being a guiding principle of the EU's 2020 strategy on smart, sustainable and inclusive growth.

This book represents an important contribution to ensuring a wide audience for knowledge sharing and the raising of awareness. The challenges of sustainability for industries, such as the cosmetics and personal care industry, need to be understood in order for sustainable practices to be included in the whole life cycle of our products. The growth in awareness and the continued commitment of the cosmetic product and personal care industry will ensure that sustainability is always embedded in the industry's culture and operations, taking care of the generations to come.

Bertil Heerink
Director-General of Cosmetics Europe – The Personal Care Association

About the Contributors

Paula Alexander is Director, Sustainable Business for BURT'S BEES® (USA). Since joining BURT'S BEES® in 2006, she has served a variety of roles in Brand Development, Marketing and Global Insights; she has been on BURT'S BEES® Leadership Team since 2011. Paula takes a whole systems approach to sustainability, focusing on source to disposal and ensures that BURT'S BEES® prioritises sustainability in all business functions at all levels. As President of The BURT'S BEES® Greater Good Foundation, she leads the giving programmes that promote community development and environmental health while reinforcing the brand mission.

Paula began her career at Unilever, where she spent eight years in brand management, working on the Dove, Suave, Lever 2000 and Snuggle brands. She holds a BBA from Texas A&M University and an MBA from Duke University's Fuqua School of Business.

Frédéric Anklin is Corporate Communications Manager at Weleda Group (Switzerland). In this position, he helps communicate his company's approach to social issues and sustainability through a variety of internal and external channels, including Weleda's annual and sustainability reports (as co-editor) and its international management newsletter (as editor-in-chief). He has eight years experience in communications and international relations. His last position was Programme Assistant for the Human Security division of the Swiss Federal Department of Foreign Affairs.

Frédéric has an MA in History, Political Science and Scandinavian Languages from the University of Basel, Switzerland. A Swiss and Norwegian national, he grew up in the Philippines and Singapore.

Judi Beerling is Technical Research Manager at Organic Monitor (UK). In this capacity, Judi provides technical consulting, advice and training on natural and organic raw material selection and product formulation techniques. She has over 30 years experience in cosmetic formulations, much of which was at Quest

International; her last position was Global Sensory Applications and Product Innovation Manager.

Judi has an MBA from the Open University Business School and is a Chartered Chemist. She is an active member of a number of industry associations. She was past President of the UK Society of Cosmetic Scientists (SCS) and was the treasurer of the International Federation of Societies of Cosmetic Chemists (IFSCC). She has also been a visiting lecturer at the University of the Arts (London College of Fashion) where she taught skin care formulation and product development.

Charles J. 'Chuck' Bennett is Vice President, Aveda Earth and Community Care (USA), responsible for leading the company in realising its commitment to the Aveda Mission. He has over 35 years of experience working in varied areas of environmental and social responsibility from teaching, research and writing, to corporate leadership in several businesses. He will be retiring from Aveda and the field in June 2013, however he looks forward to being engaged in meeting sustainability challenges.

Chuck received MA and PhD degrees in Geography from Syracuse University and an undergraduate degree from Middlebury College. He has taught in the State University of New York, written extensively about sustainability issues for The Conference Board in New York and held lead sustainability roles at Adolph Coors Company and Nabisco Foods. Married for over 40 years, Chuck has two adult children and four grandchildren who will become an increasing focus of his life.

David Bronner is President of Dr. Bronner's Magic Soaps (USA). The company is based in Escondido, California, where it is run by him and his brother Mike Bronner. It was established in 1948 by their grandfather, Dr. Emanuel Bronner, who was a third-generation master soap maker from a Jewish soap making family in Germany that began production in 1858.

Dr. Bronner's makes soaps and personal care products certified under the USDA National Organic Program. All major ingredients are from certified Fair Trade sources, such as coconut oil and palm oil. The company's fair trade organic olive oil is sourced from projects in Israel and Palestine as testament to Dr. Bronner's vision of peace, as reflected on all product labels. David runs the company much like a charitable engine in honour of his grandfather and father's ideals. A share of profits are donated to worthwhile causes and charities. Areas of particular focus are youth programmes, fair trade and organic agriculture and drug policy reform.

Michael S. Brown is a partner in the sustainability consulting firm, Brown and Wilmanns Environmental, LLC (USA). At this company, he helps clients with sustainability strategy, goals and metrics, benchmarking and multi-dimensional life-cycle evaluations of materials and products, including the Nike Materials Sustainability Index. Mike's 30 years' experience includes establishing the internal environmental program at Patagonia where he led a number of state-of-the-art initiatives.

Mike received MRP and PhD degrees from Cornell University, specialising in environmental management and health policy. He has taught Industrial Ecology at UC Santa Barbara. He is co-author of the book *Workers at Risk* and serves as an Assistant Editor for the *Journal of Industrial Ecology* and is a Public Member of the California Ocean Protection Council. He is married and has three kids scattered across the globe with whom he occasionally gets to share riding waves.

Jean-Florent Campion is Sustainable Development Manager in the Research and Innovation division of L'Oréal (France). In this capacity, he is in charge of products' eco-conception and international coordination, as well as the environmental footprint experiment project from the French initiative 'Grenelle de l'Environnement'.

Prior to joining L'Oréal, he worked at Rhone-Poulenc and Alcatel. At the former company, he developed Sol-gel processes for the semiconductor industry at the Beckman Institute for Advanced Science in Champaign-Urbana Illinois (USA). At the Alcatel-Alstom Research Center (France), he co-developed various processes for optical fibre manufacturing; he is a co-inventor of one of the two optical fibre manufacturing processes: Advanced Plasma Vapor Deposition process (APVD). Jean-Florent holds 21 patents and has written 8 research papers. He is also associated professor in environmental law at the University of Paris 13 at the Galileo Institute. He also has a Master in chemistry and a Master in Law in intellectual property.

John A. Delfausse is the principle of Sustainable Packaging Solutions, LLC (USA). He was previously Vice President – Global Package Development, and Chief Environmental Officer for Estee Lauder Corporate Packaging. In this role, he was responsible for implementing the company's commitment to responsible packaging. Under his leadership, Aveda (an Estee Lauder Company) won several packaging awards, including the HBA 2003 International Package Design Award, the CPC Editor's Choice Award, and both the Ameristar Award and 3M Integrity Award for the 2006 Earth Month Candle packaging. John was honoured in 2008 as one of Packaging Strategies' Most Influentials in Packaging.

John Delfausse is a founding member of the Sustainable Packaging Coalition. He is also as a member of the IOPP Packaging Executive Council, the Packaging World Sustainable Advisory Board and the FIT Packaging Design Department Industry Advisory Board. In 2011, John founded the Sustainable Packaging Cosmetic Roundtable in order to focus the industry on common opportunities.

Darrin C. Duber-Smith, MS, MBA, is president of Green Marketing, Inc (USA). He has over 25 years of expertise in the marketing and management profession, including extensive experience with natural, organic, and green/sustainable goods and services. He is a co-founder of the Lifestyles of Health and Sustainability (LOHAS) market concept and leader of the first US industry task force that helped frame an industry definition of 'natural'. He has published over 60 articles in trade publications and has presented at scores of executive-level events.

Darrin is a Marketing Professor at the Metropolitan State University of Denver School of Business as well as the Affiliate Marketing Faculty at the Leeds School of Business at the University of Colorado-Boulder. He was the recipient of the *Wall Street Journal*'s In-Education Distinguished Professor Award for 2009, and is author of Cengage Learning's Know Now Marketing blog.

Eduardo Escobedo is Director of Responsible Ecosystems Sourcing Platform, RESP (Switzerland). In this role, he supports the cosmetics, fashion and jewellery industries in promoting the sustainable use of natural resources and conservation of biodiversity through enhanced business models and tools.

Over the past 15 years, Eduardo has held a number of expert positions in international organisations, government and companies. Prior to founding RESP, Eduardo worked at the United Nations Conference on Trade and Development (UNCTAD) in the area of sustainable trade of biodiversity-based goods. He has also worked for the Mexican government as trade negotiator at the World Trade Organisation and the European Union. Eduardo holds a BSc in Industrial and Systems Engineering from the Instituto Tecnologico de Monterrey (ITESM) and an MA in Development Studies from the University of Geneva. He also holds post-graduate diplomas from the University of Vienna, the University of Geneva and IESE Business School.

Anne van Haeften is Brand Design Director at the multi-disciplinary design agency Reggs (The Netherlands). In this capacity, she translates a brand or business strategy into a design strategy. Anne has a Bachelor degree in Corporate Communications. She is also board member of Professional Passionates, a foundation which connects individual professionals and provides them hands-on tools to give them power to help their organisation change in a sustainable way. She is also communication manager at the ProPortion Foundation, a foundation which supports organisations in realising social business that serves low-income consumer groups in developing countries.

Bertil Heerink is Director-General of Cosmetics Europe – The Personal Care Association (Belgium), the voice of Europe's EUR 80 billion cosmetics industry. Mr Heerink is a Dutch national and holds a degree in Modern History and International Law from the University of Leiden. He spent a large part of his career in the Dutch civil service. Among other responsibilities, he was head of EU affairs in the Netherlands Ministry for the Environment, Counsellor of Embassy at the Netherlands Permanent Representation to the European Union and diplomatic adviser to the European Commission on multilateral negotiations in the field of environment. Before joining Cosmetics Europe (previously Colipa) as Director General in 2002, he was Director EU Government Affairs at the European Chemical Industry Council (Cefic) in Brussels.

Lance Kalish is the co-founder and former COO of San Francisco-based Yes To Inc.™ (makers of Yes To Carrots, Yes To Cucumbers, Yes To Tomatoes, Yes To Blueberries and Yes To Grapefruit). Lance is a serial entrepreneur and has been involved in numerous other start-ups in the USA and Australia; he operates in the capacity of board member, investor, shareholder, advisor and mentor. Lance also co-founded Trendtrade International, a brand management company involved in international business development, export management, distribution deal negotiation and facilitation. Lance has previously worked as a financial analyst in the Mergers and Acquisitions division of PricewaterhouseCoopers.

Giorgos (George) Korres is the CEO and Founder of Korres Natural Products (Greece). Whilst studying Pharmacology at the University of Athens in 1988, Giorgos started working at Greece's oldest herbal pharmacy. In 1992, he bought the pharmacy and within five years, he had set up a herbal remedy production lab with the National Organisation for Medicines' authorisation. Having developed an intimate knowledge of natural ingredients and their application in over 3000 herbal remedies, Giorgos created the Korres company in 1996. The aim was to produce safe, clinically effective and affordable natural skincare products of interesting design. Today, his company employs over 250 people and its portfolio includes over 400 natural and/or certified organic products.

Giorgos is involved in brand development and research in collaboration with the National Hellenic Research Foundation, Athens University and various worldwide research centres. Korres products utilise clinically efficacious, endemic Greek herbs, whilst cooperating with local communities, agricultural unions and organic farmers.

Ido Leffler is Co-Founder and 'Chief Carrot Lover' of San Francisco-based Yes To Inc. (makers of Yes To Carrots, Yes To Cucumbers, Yes To Tomatoes, Yes To Blueberries and Yes To Grapefruit). Since 2006, Leffler and his team have built Yes To™ into an international natural beauty brand, with distribution in over 26 000 stores in 25 countries (including retailers such as Target, Wal-Mart, Walgreens, Whole Foods Market and Sephora Europe). Ido is currently on the United Nations Foundation Global Entrepreneur Council and sits on numerous corporate boards and corporate advisory boards

Rik Kutsch Lojenga is the Executive Director of the Union for Ethical BioTrade (The Netherlands). Prior to forming UEBT, he worked for over 10 years at the United Nations Conference on Trade and Development (UNCTAD) where he was involved in the BioTrade Initiative. Rik is also President of the Board of the ISEAL

Alliance, an umbrella of sustainability standards of which UEBT is member. Rik has a Master's degree in business and environmental economics.

Ramani Narayan is University Distinguished Professor in the Department of Chemical Engineering and Materials Science at Michigan State University (USA). He has 148 refereed publications, 28 patents and is a successful entrepreneur with commercialised technologies in the bioplastics space. He has won many awards – named MSU University Distinguished Professor in 2007 – the highest honour that can be bestowed on a faculty member by the university; Governor's Award for commercialization excellence; University Distinguished Faculty Award, Withrow Distinguished Scholar award; Fulbright Distinguished Lectureship Chair at University of Lisbon, Portugal; ASTM International Award of merit and named ASTM Fellow. He is the USA's technical expert, Chairman, and Convener in several ISO standards committees, as well as the Chair of the ASTM committee on biobased products and environmentally degradable plastics (D20.96); Scientific Chair of the Biodegradable Products Institute (BPI), North America and serves on the Board of Directors of several publicly traded companies.

Remi Pulverail is Global Head of Naturals Purchasing at Givaudan (Switzerland). Remi's passion for travel, spices and perfume were cemented in Morocco during childhood. His career in the fragrance industry started with Charabot in Grasse before joining Givaudan, where he quickly focused on the procurement of natural materials.

Remi is currently responsible for sourcing the natural ingredients used by Givaudan's Fragrance and Flavour Divisions globally. Passionate about the quality of materials and serious about the responsibility of companies towards supply chains, Remi is the driving force behind the company's ethical sourcing initiatives. He and his team currently manage more than 10 programmes worldwide, with more in development.

Mason W. Rubin is a senior at the Leeds School of Business at the University of Colorado at Boulder (USA). His emphasis of study is marketing; he is also is working towards a minor in philosophy and ethics, as well as a certificate programme in international business. Mason plans to develop a career in brand management, consumer behaviour, corporate social responsibility and/or sustainability. He has led numerous marketing-based team projects in school, as well as interned for a small company running social media-based marketing.

Amarjit Sahota is the founder and president of Organic Monitor (UK), a specialist research, consulting and training firm that specialises on the global organic and related product industries. His company hosts sustainability summits across the globe, including editions of the Sustainable Cosmetics Summit in Europe, North America, Latin America and the Asia-Pacific.

Amarjit has been tracking health, wellness and sustainable product industries for over 15 years. He has guided a wide range of clients in realising their business potential in these industries. His clients range from multi-nationals, government organisations, to dedicated sustainable product companies. Amarjit has degrees in Applied Chemistry and Management Science, as well as a Postgraduate Diploma in Marketing. Apart from editing this book, he has co-authored a number of books and various articles on sustainability and ecological products.

Filipe Tomazelli Sabará is Business Director at Beraca (Brazil). In this position, he is in charge of business development and has managerial responsibilities in the family business. His career at Beraca started in 2002 as Marketing Manager of its six business units in Brazil. Since 2007, Filipe has been responsible for Beraca's Global Business Development in more than 40 countries; he is also responsible for the company's European operations. Filipe has a degree in International Business and Economy from the Fundação Armando Alvares Penteado (FAAP) in São Paulo, Brazil. He also has a postgraduate in Exports and Marketing from Ibero American University in São Paulo.

Bas Schneiders has been responsible for Corporate Sustainability and Strategic Sourcing at the Weleda Group (Switzerland) since 2010. Prior to this, he was the managing director of Weleda Naturals GmbH, a 100% subsidiary of Weleda AG. He was responsible for the cultivation and production of cosmetic and pharmaceutical plants extracts and tinctures, as well as managing the global sourcing activities.

Bas has extensive experience in ethically sourced raw materials and packaging. Prior to joining Weleda, he held management positions at Tradin Organic Agriculture BV and at Eosta BV. The two Dutch companies are leading 'integrated' trading houses of organic raw materials and foodstuffs. Bas holds degrees in Business Management from the International University of Hospitality Management in Den Haag and the VU University Amsterdam.

Kathy Sheehan is Executive Vice President and General Manager of GfK Consumer Trends, a unit of GfK Custom Research North America (USA). In this role, Kathy oversees GfK Consumer Trend's premier consumer trends services – Roper Reports®, Roper Reports Worldwide, and the GfK Roper Green Gauge® study on consumers and the environment.

Kathy has been an adjunct professor in the Marketing department at LIM College in New York City and has also been a guest lecturer in the Masters in Cosmetics and Fragrance Marketing programme at the Fashion Institute of Technology (FIT) as well as the MBA programme at Baruch College. She received her MBA in

International Business and Marketing from Baruch College and her MA in International Political Economy from New York University. Kathy also holds a BA from Rutgers University with a double major in Political Science and English and a Russian language minor.

Jody Villecco is responsible for researching and maintaining the Quality Standards at Whole Foods Market (USA). In this role, her duties include developing and communicating the Company's Quality Standards for dietary supplements, personal care products and household cleaning products, and keeping the company current with technical and regulatory issues. She joined Whole Foods Market in 1997, working for the dietary supplement arm, Amrion, where she formulated nutritional supplement products, compiled research summaries and conducted nationwide training sessions on nutrition and herbs. Jody received her degree in nutritional sciences from Cornell University in 1994 and a Master's degree in nutritional anthropology from the University of Colorado in 2012. She has appeared on national television and has done many interviews with top-tier publications on nutrition and natural beauty topics.

Xavier Vital is Environmental Sustainability Services Manager at SGS North America (USA). He is a senior expert in ecodesign and life cycle assessment and sustainable supply chains, providing support to manufacturers, retailers and public bodies interested in improving their environmental footprint. Xavier manages the development of Ecodesign services for SGS North America. He has conducted, managed or reviewed around 800 simplified and complete LCAs of products or services including major projects in the cosmetic industry.

Xavier has a Masters in Ecodesign from the Open University of Cergy-Pontoise (France). He has provided his expertise to the Environmental Labelling of Consumer Products platform led by the ADEME (French Environmental Protection Agency) for the development of a methodology and Product Category Rules compliant with the ISO 14040 standard. He is also involved in the development of sustainable suppliers programmes (GSCP).

1

Introduction to Sustainability

Amarjit Sahota

1.1 INTRODUCTION TO BOOK

The world we live in is changing. The global population has reached 7 billion and is projected to reach 9 billion by 2050. There is a general agreement that the planet's resources cannot cope with such a rise in human population, especially at existing consumption rates. Human activity has contributed to, or been responsible for, climate change, loss of biodiversity, destruction of habitat for many species and other such environmental damage. Human behaviour and consumption patterns need to change if the planet is to adequately feed another 2 billion mouths.

Consumerism is also changing. As consumers become more informed, they are demanding more from the products they buy. Rising education levels, the Internet and growing use of mobile devices are making consumers more informed than at any other time in history. They are questioning product origins, production methods and ecological implications, as well as safety issues. This rise in ethical consumerism is having a major impact on the cosmetics industry. Cosmetic and ingredient companies are increasingly scrutinised by retailers and NGOs looking to safeguard consumer interests.

It is against this backdrop that the idea of this book came about. With growing scarcity of resources and rising ethical consumerism, how can the cosmetics industry become more sustainable? What are the best practices in sustainable development? What areas are cosmetic companies focusing on, and what areas need to be

Sustainability: How the Cosmetics Industry is Greening up, First Edition. Edited by Amarjit Sahota.

improved? This book aims to address such questions. Written by the industry for the industry, it should be considered a practical guide for organisations looking to make a difference in terms of sustainability.

The first chapter serves to give an introduction to the book. A brief introduction is given to sustainability, specifically the relevance of environmental and social impacts. The proceeding chapters are written by industry professionals who share their expertise in specific areas of sustainability. The final chapter (future outlook) summarises some of the key findings from the book and gives future projections.

1.2 INTRODUCTION TO SUSTAINABILITY

Sustainability has many interpretations. A widely accepted definition is that of the Brundtland Commission of the United Nations which refers to sustainability in the concept of sustainable development: 'meeting the needs of the present without compromising the ability of future generations to meet their own needs' [1].

Sustainability has three pillars or dimensions, as depicted in Figure 1.1.

I Environmental dimension – requires that natural capital remains intact. This means that the source and sink functions of the environment should not be degraded. Thus, the extraction of renewable resources should not exceed the rate at which they are renewed, and the absorptive capacity of the environment to assimilate wastes should not be exceeded. Furthermore, the extraction of non-renewable resources should be minimised and should not exceed agreed minimum strategic levels.

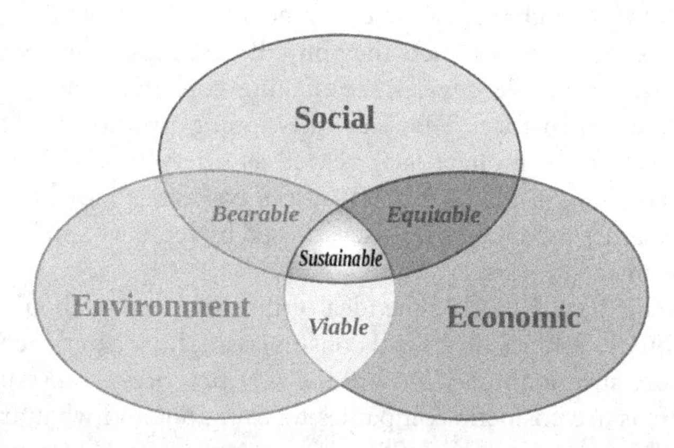

Figure 1.1 The three pillars of sustainability.
Source: (Reproduced from http://en.wikipedia.org/wiki/File:Sustainable_development.svg, authored by Johann Dréo, inspired from image 'Sustainable Development'.).

II Social dimension – requires that the cohesion of society and its ability to work towards common goals be maintained. Individual needs, such as those for health and well-being, nutrition, shelter, education and cultural expression should be met.

III Economic dimension – occurs when development, which moves towards social and environmental sustainability, is financially feasible.

These three pillars are commonly referred to as the 'triple bottom line'. They form the basis of many sustainability standards and certification systems that include the Rainforest Alliance, Fairtrade and UTZ Certified. Approaches to sustainability are discussed in the context of these three pillars throughout this book (see Section 1.6).

Although the understanding of sustainability has increased significantly in recent years, it remains an industry term. Consumers seldom use it and many are unsure over what it means. Research by the Hartman Group finds that although more than half of American consumers are familiar with the term 'sustainability', most cannot state what it means. The research polled 1606 American consumers to gain an understanding into consumer behaviour towards sustainability practices and products. Only 5% of consumers could name companies that support sustainability values, whereas 12% stated they knew where to buy such products [2].

1.3 ETHICS IN THE COSMETICS INDUSTRY

Ethics are considered very important to the cosmetics industry for it comes under closer scrutiny then other industries. Cosmetic products have traditionally been perceived as **'vanity' products**; they can be considered non-essentials mainly bought by people to improve their appearance. The industry and its business practices are therefore much more scrutinised then other related industries, such as foods, pharmaceuticals, home cleaning products and so on.

The problem with this argument is that cosmetic products are not just mascara, lipsticks and foundations; they also refer to cleansing products required for basic hygiene, like shampoos, soaps, toothpaste and shower gels, as well as deodorants, shaving creams and and moisturisers. Also included are products with specific health purposes, for example anti-acne creams, anti-inflammatory lotions, anti-lice shampoos and healing creams. In this respect, cosmetic products do not serve just the vanity needs of consumers but also the basic needs of hygiene and cleanliness.

The cosmetics industry is often targeted by the media and NGOs because of the use of **animal-testing methods**. Cosmetic companies have historically tested ingredients, as well as finished products, on animals (typically rabbits and mice) to check safety levels. Although there is a move to phase out animal testing in the industry, it is likely to be many years (if not decades) before a global ban is introduced and then enforced. Israel introduced its ban in January 2013, the EU ban is scheduled for March 2013, whilst other countries have yet to introduce regulations to ban such methods.

The UK organisation, Cruelty-Free International, is spearheading a global campaign to end animal testing for cosmetic products and ingredients. Products can be certified that they are not tested on animals; such products meet the Humane Cosmetics standard and carry the Leaping Bunny logo. The animal rights group People for the Ethical Treatment of Animals (PETA) is also campaigning to ban animal testing for cosmetic products. It has undertaken a number of advertisements using celebrities, such as ex-Pussy Doll pop singer Kimberly Wyatt (as shown in Figure 1.2).

Consumers place great emphasis on buying cosmetics with a conscience. A survey by the American online retailer Vitacost.com in October 2012 found that 75% of women prefer to purchase cosmetic products with a 'cruelty-free' logo over products without.

Since cosmetics are made up of chemicals, the industry is closely linked to the **chemical industry**. Indeed, many of the largest chemical companies in the world supply speciality chemicals to the cosmetics industry; such companies include BASF, Dow Chemical, Evonik, Rhodia and Eastman Chemical. Some of the unethical business practices of the chemical industry, like environmental pollution, also become associated with the cosmetics industry.

Cosmetic companies are also coming under the microscope for natural **ingredient sourcing**. The industry is one of the largest users of palm oil, a vegetable oil that is predominantly grown in Indonesia and Malaysia. Unethical sourcing of palm oil has been responsible for the destruction of tropical rainforests, threatening the habitat of endangered orang-utans. Unilever, one of the largest cosmetic companies in the world, was named as a buyer of unethical palm oil by Greenpeace in November 2009. The move led Unilever to drop its Indonesian supplier and make a commitment to only source sustainable palm oil certified by the Roundtable of Sustainable Palm Oil (RSPO).

The environmental damage caused by cosmetic **finished products** is also coming under the spotlight. A number of studies have reported on the adverse effects of cosmetic ingredients on the environment. In August 2012, research by Arizona State University and federal authorities found Minnesota waterways to be contaminated by cosmetic ingredients. Anti-microbial ingredients like triclocarban and triclosan are present in soaps, disinfectants and sanitisers; they are getting into fresh waterways from waste treatment plants after entering sewers from consumer households. Apart from their endocrine-disrupting abilities, anti-microbial chemicals are toxic to aquatic bacteria. Tricolsan also prohibits photosynthesis in diatom algae, responsible for a large part of the photosynthesis on Earth.

Microplastics in formulations are also accumulating in the seas and oceans, disrupting marine ecosystems. Micro beads are used in soaps, scrubs and shower

Figure 1.2 Advertising campaign by PETA on animal testing for cosmetics.
Source: PETA: Image: © karlgrant.com.

gels for exfoliating and texturising purposes. Since they are slow to biodegrade, they accumulate in water and are ingested by marine life and create damage.

Cosmetic packaging has a detrimental effect on the environment. Luxury skin care products and perfumes are typically housed in layers of packaging. The use of such excess packaging is often questioned, considering that resources are becoming scarce and they contribute to high product prices. A larger concern is the packaging impact on the environment. Plastics are the most widely used packaging material in the cosmetics industry; popular because of their flexibility and light weight; however, they create environmental pollution and do not biodegrade in landfills.

Basic hygiene personal care products create as much environmental pollution as other products. For instance, about 23 000 tonnes of toothbrushes end up in landfill in the USA each year, whilst about 2 billion disposable razors are thrown away. Cosmetic packaging also contributes to marine pollution. Plastic packaging in the sea has been linked to the injury and death of seagulls, fish and other marine life.

The **safety of cosmetic ingredients** is also the subject of much attention. Some studies suggest that phthalates – widely used as solvents in hair sprays, nail varnishes and perfumes – act as potential endocrine disruptors. Parabens, a family of chemical preservatives present in thousands of cosmetic products, are thought to mimic oestrogen and are linked to breast cancer. Other cosmetic chemicals that are linked to health conditions include aluminium salts, petrochemicals oils, triclosan, formaldehyde, mercury and other heavy metals. It is worth noting that many of these cosmetic chemicals may be associated with health risks, however scientific evidence is often lacking. Consumer perception is often stronger than reality when it comes to product safety.

A major issue about the safety of such chemicals is the variation in regulations between different regions and countries. For instance, the EU banned the use of phthalates in 2003, however it is still permitted in other regions. Some countries have lax enforcement of regulations, leading to potentially serious incidents. In December 2012, the interest group EcoWaste Coalition found that mercury-laden cosmetics were being sold in the Philippines. Although the national government had banned the sale of cosmetics with mercury because of health risks, many retailers were ignoring the ban.

In summary, the cosmetics industry gets more then its fair share of scrutiny because of the perceived nature of cosmetic products. Apart from the ethical issues surrounding animal testing, the industry is often criticised for its selection and use of raw materials, environmental impacts and safety issues of finished products.

1.4 DRIVERS OF SUSTAINABILITY

Whereas the previous section highlighted the reason why the cosmetics industry faces close scrutiny, this section states the factors pushing the sustainability agenda in the industry.

1.4.1 Rise in Ethical Consumerism

As stated in the introductory pages, consumers are increasingly asking ethical and environmental questions before purchasing products. The growing number of media reports on sustainability, provenance and the environment is raising consumer awareness of environmental and social issues. Consumers are realising that their purchasing decisions are having a direct impact on the environment and social communities. Furthermore, consumer demand for organic and natural foods is spilling over onto non-food products. They are looking for beauty products that are made ethically and contain natural and organic ingredients. The high growth in the natural and organic cosmetics market is testament to this; global sales increased from less then $1 billion in the mid 1990s to $9.1 billion in 2011 [3]. Consumers are also looking at the carbon footprint and ingredient sources of beauty products.

This change in consumer behaviour is affecting demand for cosmetics and personal care products. Although the primary decision for cosmetic products remains personal (meeting health and wellness needs), ethical and environmental considerations are becoming increasingly important when consumers buy these products.

1.4.2 Pressure from the Media and NGOs

The media is playing an important role in raising the profile of environmental, social and sustainability issues. As consumers become more aware of these issues, they are questioning corporate ethics and demanding products that meet high ethical and ecological standards. In some cases, these demands translate into boycotting of products that have a detrimental impact on the environment and social communities.

Non-Governmental Organisations (NGOs) and industry groups are also playing a key role in raising public awareness. For instance, Greenpeace France released the Cosmetox Guide in 2005. The report gives details of synthetic chemicals and their potential toxicity in beauty products. The guide classifies beauty products as green, orange or red according to information provided on the chemical composition of their products. Many L'Oréal products were given a red rating in the report.

In the USA, the Campaign for Safe Cosmetics is putting pressure on the cosmetics industry to remove potentially harmful chemicals. In August 2012, Johnson & Johnson bowed to such pressure to remove potentially carcinogenic chemicals, like phthalates and triclosan, from its product range by 2015. A year earlier, it had agreed to remove 1,4 dioxane and formaldehydes from its formulations by 2014.

Greenpeace also published a report titled 'How the Palm Oil Industry is Cooking the Climate' in 2007. The report implicated personal care companies, such as Procter & Gamble and Unilever, for the purchase of palm oil from suppliers that are actively engaged in burning forests and draining peat lands in Indonesia. In 2008, Greenpeace issued a follow-up report, 'How Unilever Palm Oil Suppliers are Burning Up Borneo' with further evidence of the expansion of the palm oil sector into rainforests, orang-utan habitat and peat lands in Kalimantan, Indonesia. The report linked the majority of the largest producers in Indonesia to Unilever, one of the largest palm oil buyers in the world.

Pressure from the media and NGOs has a significant impact on consumers and cosmetic manufacturers. For instance, the Greenpeace reports on palm oil led many cosmetic companies to take corrective measures. Unilever suspended its US $33 million contract with the Indonesian firm, Sinar Mas, in 2009 after an independent audit proved that Sinar Mas was involved in the destruction of rainforests. Unilever also blacklisted PT Duta Palma and declared that it will only use palm oil from sustainable sources by 2015.

1.4.3 Environmental Changes and Finite Resources

A number of external factors are encouraging cosmetic companies to adopt sustainability initiatives. As the global population increases and life expectancy rises, there is growing pressure on the planet's finite natural resources. Shortages of energy, water and other raw materials, as well as concerns about climate change, are making organisations focus on efficiency and sustainability issues, such as ethical sourcing and waste management.

Resource reduction is a major feature of the sustainability plans of large cosmetic companies. For instance, Procter & Gamble stated in 2007 that it plans to reduce its energy consumption, water consumption, disposed waste and carbon emissions by 20% by 2020.

Cosmetic and ingredient companies are also developing green formulations because of the dwindling supply of petrochemical feedstock. With global oil supply peaking, the cosmetic industry is being forced to consider alternative feedstock for ingredients. This development is leading to new green cosmetic ingredients emerging from agricultural-based raw materials. For instance, the chemical company, Rhodia, launched a range of hair care polymers made from vegetable polymers in April 2010.

The company Du Pont Tate & Lyle Products was formed in 2007 to capitalise on this trend. The enterprise is a joint venture between the British sugar company Tate & Lyle and the American chemicals giant Du Pont. The enterprise has used the expertise of its parent companies to produce a novel range of natural ingredients that are synthesised from sugar. Heliae is another company that is developing similar green ingredients from sea algae.

1.4.4 Pressure from the Supply Chain

Sustainability initiatives are also encouraged by operators in the supply chain. NGOs and rising consumer expectations are causing retailers to put pressure on suppliers. Retailers are increasingly asking for traceability and transparency from their suppliers, especially in terms of ingredients, adherence to labour laws and treaties and production methods. For instance, the UK retailer Marks & Spencer has put pressure on its suppliers to switch to sustainable palm oil so it can meet its target of 100% sustainable palm by 2015.

Wal-Mart has an ambitious plan to have sustainability measures for each product it sells. By encouraging sustainability across its supply chains, Wal-Mart believes it can reduce 20 million tonnes of greenhouse gas emissions by 2015. This is the equivalent to taking 3.8 million cars off the road in a year.

1.4.5 Laws and Regulation

Environmental law is commonly referred to as a complex and interlocking body of treaties, conventions, statutes, regulations and common law that, very broadly, operates to regulate the interaction of humanity and the rest of the bio-physical or natural environment, toward the purpose of reducing the impacts of human activity, both on the natural environment and on humanity itself.

Pollution does not respect political boundaries, making international law an important aspect of environmental law. A plethora of legally binding international agreements now encompass a wide variety of issues, ranging from terrestrial, marine and atmospheric pollution to wildlife and biodiversity protection. About 1000 environmental law treaties exist; no other area of law has generated such a large body of conventions on a specific topic. Many protocols have been built from these treaties. The Kyoto Protocol is the most widely known, aimed at combating global warming.

Providing the environmental impact of a product in the form of carbon footprint and water consumption labels is becoming popular. In the USA, Wal-Mart requests some of its suppliers to declare certain environment indicators. In the UK, the Carbon Trust has developed a carbon footprint scheme that gives a product's carbon usage. Such schemes remain voluntary; however, they could become mandatory in certain countries.

A growing number of countries are introducing regulations covering certain aspects of sustainability. The French government introduced the Grenelle Law for the environment, which plans to make the labelling of a number of environmental indicators on certain products compulsory. In Brazil, the government has a biodiversity regulation that penalises companies for not sharing the benefits of novel ingredients with local communities [4].

Corporate social responsibility (CSR) and sustainability reporting is also becoming compulsory in certain countries. Governments are forcing publicly listed companies or state-owned enterprises to provide sustainability reports to improve transparency and corporate accountability. Such countries include the UK, France, Denmark, Sweden, Brazil and South Africa. Similar policies are being developed in China and other countries to encourage sustainability reporting.

1.4.6 Business Benefits

There is a growing realisation that sustainability can bring many business benefits. By focusing companies on becoming more efficient in terms of resource use, sustainability can have a direct impact on the bottom line. Reducing raw material and energy usage, as well as waste materials, can lower business costs. Thus, there is greater profitability by 'doing more with less'.

Sustainability can also improve company morale and performance. Employees are generally more motivated if they are contributing to 'good causes', resulting in higher performance levels and greater staff longevity. It can also improve/strengthen customer loyalty and create brand differentiation.

Sustainability also plays an important role in managing risk, which can reduce the likelihood of reputation harm, cost of disputes and litigation, as well as other detrimental situations. Some companies have been able to measure the benefits of sustainability initiatives; for instance, the UK retailer Marks & Spencer stated that its Plan A (sustainability plan) created an additional GB £185 million brand benefit between 2007 and 2012 [5].

1.5 SUSTAINABILITY REPORTING

A key challenge for many companies involved in sustainability is to measure and report their actions to stakeholders. Excess communications, especially in marketing, can lead to accusations of greenwashing. At the other extreme, lack of reporting or communicating can create indifference from stakeholders. This section looks at the two common ways companies report their sustainability actions.

1.5.1 CSR and Sustainability Reports

It is becoming increasingly common for large corporations, especially publicly listed ones, to produce CSR or sustainability reports. Indeed, over 80% of Global Fortune 250 companies (G250) now disclose their sustainability performance in sustainability or corporate social responsibility reports. All cosmetic companies in the G500, including Procter & Gamble, Unilever, Johnson & Johnson, Henkel and L'Oréal, published CSR reports in 2012. Smaller companies are following in their footsteps: Burt's Bees and Weleda are two natural cosmetic companies that have started producing such reports.

One reason for the rise in CSR and sustainability reporting is legislation. As stated in the previous section, a growing number of countries are making it compulsory for companies to produce such reports.

Corporate disclosure on economic, environmental and social performance is fast becoming a mainstream activity, fuelled by public demand and corporate peer pressure. New research suggests that many companies are no longer just issuing a single paper report but communicating their sustainability performance across a multitude of communications channels.

Whilst there is no universal format yet, there are a wide number of standards available for CSR and sustainability reports, such as those from the Global Reporting Initiative (GRI), Business in the Community (BITC) and the World Business Council for Sustainable Development (WBCSD). While each of these standards has merit, GRI's Sustainability Reporting Framework is the most commonly used. That framework – of which the Sustainability Reporting Guidelines are the cornerstone – is said to be universally applicable, regardless of the size, sector or location of an organisation. The guidelines contain principles and guidance as well as standard disclosures – including indicators – to outline a disclosure framework that organisations can voluntarily, flexibly and incrementally adopt. The framework is developed by a large multi-stakeholder network of thousands of experts from over 60 countries, and is continuously improved and expanded. The core guidelines are now in their third generation (G3); they were released in October 2006 following a three-year development period that engaged more than 3000 individuals from diverse sectors across the globe. The fourth generation (G4) of guidelines are expected in May 2013. Sector Supplements are also available to complement the use of the core guidelines and respond to the limits of a one-size-fits-all approach, but there is no specific supplement for the cosmetic industry.

1.5.2 Communicating to Consumers

Since sustainability and CSR reports are seldom read by consumers, a major challenge for cosmetic companies is how to get their sustainability message across to consumers.

The most successful communication programmes are those that spread the word about sustainability related efforts through multiple channels, position the issues in a relevant way to each stakeholder and extend messages beyond traditional media relations to include all forms of online marketing, grassroots and viral marketing, product packaging, special events, conferences and seminars, public service advertising (PSA) and advertising and expert spokespersons, among others. Direct communication at the point-of-sale, eco-labels on products, engaging consumers through campaigns and efficient advertising are just a few examples of winning strategies [2].

An example of point-of-sale communications is by the American natural cosmetics company, Kiss My Face. It is demonstrating its eco-consciousness with Greenhouse, an in-store point of purchase display that is not only recyclable, but made from recycled materials. To make consumers aware of its efforts, Kiss My Face provides the following explanation on the display: 'Because you care, this display is planet friendly. To minimize waste and maximize recycled and sustainable resources, all corrugated components are from minimum 90% recycled materials. Graphics are printed with non-polluting water solvent inks. Steel parts are made from 70% recycled material and powder coating on metal parts use non-VOC materials that eliminate airborne pollutants. Wood components are biodegradable and treated with an environmentally safe lacquer.'

Aveda created a popular public outreach programme to recycle polypropylene bottle caps. Aveda's 'Recycle Caps with Aveda' campaign has become a popular and effective marketing technique to show the company's concern for the environment, whilst partnering with schools and environmentally conscious consumers. More details are given in Chapter 6.

Burt's Bees concentrates its efforts on consumer education. One of its advertising campaigns focused on beauty benefits of truly natural products in an atypical way. The company did not make overt environmental claims but targeted ingredients, not brands, with suspected human health risks (petrolatum, dimethicone, DMDM hydantoin, sodium lauryl sulfate, parabens). For example, the message on the ad for its Baby Bee Buttermilk Lotion was: 'How do you get all the snugly without the scary? Buttermilk vs. Parabens [. . .] Have you read your baby lotion label lately?'

Social media provides a powerful channel for cosmetic companies to not only communicate their green credentials, but also strengthen customer relationships. Social media enables brands to communicate directly with their customers, facilitating long-term relationships. They also force brands to be transparent in their communications with consumers since there is open dialogue. This is an area where sustainability communications is expected to rise in the cosmetics industry in the coming years [5].

The subject of green marketing is covered in detail in Chapter 11, whilst case studies of green brands undertaking marketing communications are given in Chapter 12.

1.6 GUIDE TO BOOK CHAPTERS

This section gives a guide to the 14 chapters in this book. The overriding purpose of this book is to outline methods of reducing environmental and social impacts of cosmetic products by highlighting industry best practices.

Figure 1.3 highlights some methods of reducing the environmental footprint of cosmetic products. The bullet points are not exhaustive, but state some of the

Ecological footprint

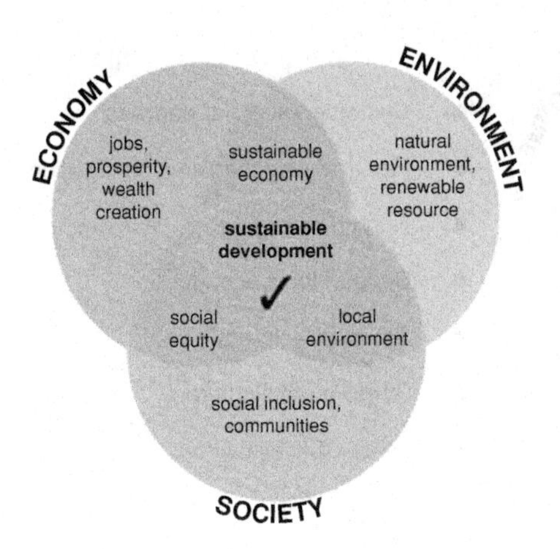

- Ethical sourcing of ingredients
- Organic and sustainable production methods
- Green chemistry / formulations
- Sustainable packaging
- Energy sources & use
- Carbon & water management
- Operational efficiency
- Waste management

Figure 1.3 The ecological aspects of sustainability.
Source: Reproduced from [4]. Sustainable Cosmetics Summit Latin America, Sao Paulo, September 24–26 2012.

common focus areas. The book has specific chapters on subject areas that are most relevant to cosmetic and ingredient companies.

Chapter 2 gives a detailed account of the environmental impact of cosmetic products, giving details of methodologies and tools. L'Oréal, the world's largest cosmetics company, shares some of its work on reducing environmental impacts in the same chapter.

Ethical sourcing is an area that cosmetic and ingredient companies have focused heavily on. **Chapter 4** gives case studies of two companies that are spearheading ethical sourcing projects. Givaudan, the world's leading fragrance and flavouring company, gives details of its Naturals Sourcing programme. Beraca, the frontrunner in ethical sourcing projects in Brazil, shares some its pioneering projects in the Amazon.

Biodiversity has been described as the greatest environmental challenge faced by the planet, after climate change. **Chapter 5** gives a detailed account of biodiversity and its sustainable use in the cosmetics industry. The contributors are from the Union for Ethical BioTrade (UEBT) and the United Nations Conference of Trade and Development (UNCTAD).

Chapter 6 has three contributions covering the growing area of sustainable packaging. The impact of design and materials is discussed by two experts, followed by a case study from Aveda, which is the leading user of recycled plastics in the cosmetics industry.

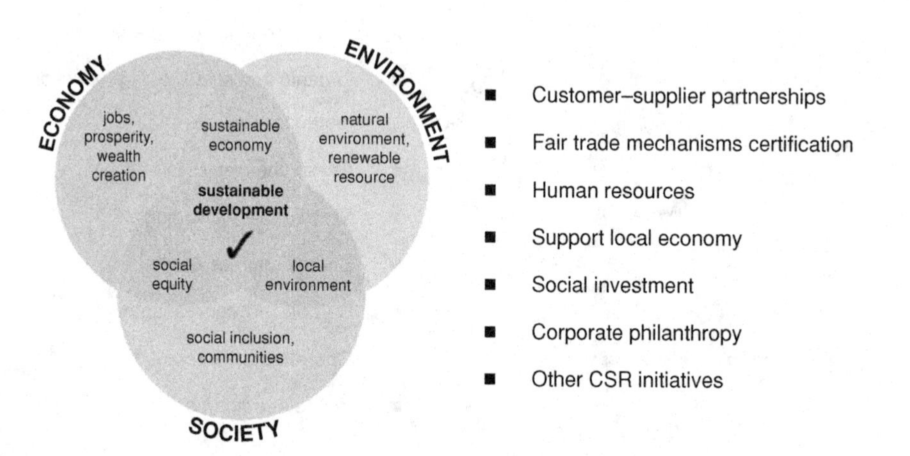

Figure 1.4 The social aspects of sustainability.
Source: Reproduced from [4]. Sustainable Cosmetics Summit Latin America, Sao Paulo, September 24–26 2012.

Another contribution from Aveda is on energy and waste management. In **Chapter 7**, the American company gives details on its renewable energy and waste management programmes. The company is the largest user of renewable energy in the cosmetics industry.

The social footprint of cosmetic products is described in detail in **Chapter 3**. The Swiss natural cosmetics firm, Weleda, states how cosmetic companies can address their social impact (popular methods are listed in Figure 1.4). In the second part of the chapter, details are given on some projects that are making a social difference. As well as this chapter, the social impacts of ethical sourcing projects are covered in Chapter 4.

The role of CSR and corporate philanthropy in the cosmetics industry are discussed in **Chapter 8**. Burt's Bees and Dr. Bronner's Magic Soaps are covered as case studies. The natural cosmetic companies have spearheaded a number of CSR/corporate philanthropy projects.

Many cosmetic companies are looking at green formulations to reduce their environmental footprints. **Chapter 9** discusses important developments in green formulations and ingredients, especially the replacement of synthetic ingredients in cosmetic products.

Chapter 10 gives an update on the growing number of green standards, certification schemes and indices in the cosmetics industry. Details are given of natural, organic and fair trade standards, as well as sustainability and corporate responsibility indexes.

Chapters 11–13 discuss marketing issues related to sustainability in the cosmetics industry. **Chapter 11** gives an introduction to green marketing issues and pitfalls, highlighting some of the ways companies are making green marketing claims.

Marketing case studies are given of green brands and a natural food retailer in **Chapter 12**. The Israeli brand Yes To and Greek brand Korres share their experiences in building distribution and marketing communications. Whole Foods Market, the world's largest natural and organic food retail chain, gives details of its Body Care Quality Standards for cosmetic and personal care products.

A consumer research agency gives some insights into the green consumer in **Chapter 13**. Some comparisons are made between green consumers in various parts of the world.

The final **Chapter 14** summarises some of the key findings in this book and provides some growth projections. What are some of the shortcomings in sustainability in the cosmetics industry and what can we expect to see in the future?

REFERENCES

[1] United Nations General Assembly Report of the World Commission on Environment and Development: Our Common Future, 1987.
[2] Organic Monitor (2010) Strategic Insights report on CSR & Sustainability in the Cosmetics Industry, London, UK.
[3] Organic Monitor (2011) The Global Market for Natural & Organic Cosmetics, London, UK.
[4] Sustainable Cosmetics Summit Latin America, Sao Paulo, September 24-26 2012.
[5] Sustainable Cosmetics Summit Europe, Paris, November 28–30 2012.

2

Environmental Impacts of Cosmetic Products

PART 1: THE GROWING IMPORTANCE OF METRICS

Xavier Vital

Sustainability metrics are becoming more and more important to business, both in conversations with stakeholders as well as within companies internally. Tracking and reporting sustainability metrics internally ensures improved performance. Tracking metrics is the first step to achieving cost savings and to significantly reducing environmental impacts.

Sustainability reports are using metrics to provide transparency and to support the business internally and externally. Metrics are required by stakeholders to support green claims and avoid any accusation of greenwashing by NGOs or other stakeholders. Without comprehensive internal reporting on metrics, companies do not have a clear picture of how they are performing over time. Metrics are required to evaluate and improve their practices.

Metrics that go beyond the carbon footprint are essential to inspire companies to accelerate along the path of sustainability. They can also help companies save money, revealing information about products, processes and systems with poor efficiency and provide direction for improvement.

The following section describes the corporate carbon footprint, usually considered by companies as a very important metric to consider.

Sustainability: How the Cosmetics Industry is Greening up, First Edition. Edited by Amarjit Sahota.

2.1.1 CORPORATE CARBON FOOTPRINTING

A corporate carbon footprint can be defined as "the total set of greenhouse gas (GHG) emissions caused by an organization."

GHGs can be emitted through transport, the production and consumption of fuels, manufacturing goods, materials, wood as well as the construction of roads and buildings, and the provision of services. For simplicity of reporting and interpretation, the emission of all GHGs is typically expressed in terms of the effective equivalent amount of carbon dioxide.

To be successful, a corporate carbon footprint system should have the following characteristics:

- Comprehensive: Incorporates scope 1, 2, and 3 emissions.
- Current: Enables updates at regular intervals and comparisons across reporting periods.
- Auditable: Traces transactions and enables independent reviews for compliance.
- Flexible: Incorporates data from onsite measurement sources to life cycle analysis.
- Standards-based: Accommodates existing generally accepted standards and emerging standards.
- Scalable: Accommodates a growing volume and complexity of business operations.
- Efficient: Delivers data in the timeframe required for decision making.

2.1.1.1 Introducing the Concept of "Scope"

Three "scopes" are defined for GHG accounting and reporting purposes by the GHG Protocol (scope 1, scope 2, and scope 3) (see Figure 2.1.1).

- **Scope 1: Direct GHG emissions**
  Direct GHG emissions occur from sources that are owned or controlled by the company, for example, emissions from fuel combustion in owned or controlled boilers, furnaces, vehicles, and so on; emissions from chemical production in owned or controlled process equipment.
- **Scope 2: Electricity indirect GHG emissions**
  Scope 2 accounts for GHG emissions from the generation of purchased electricity consumed by the company. Purchased electricity is defined as electricity that is purchased or otherwise brought into the organizational boundary of the company. However, Scope 2 emissions physically occur at the facility where electricity is generated.

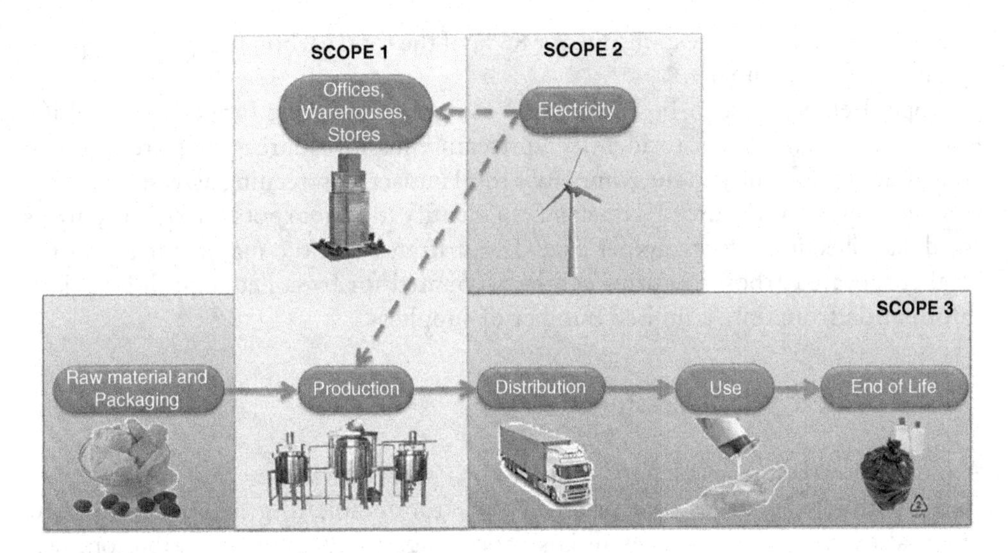

Figure 2.1.1 The scopes of GHG accounting and reporting.
Source: SGS.

- **Scope 3: Other indirect GHG emissions**
  Scope 3 is an optional reporting category that allows for the treatment of all other indirect emissions. Scope 3 emissions are a consequence of the activities of the company, but occur from sources not owned or controlled by the company. Some examples of scope 3 activities are extraction and production of purchased raw materials; transportation of purchased fuels; the transportation, use, and disposal of sold products.

2.1.1.1.1 Limitations

Although the data provided on carbon accounting ledgers may cover onsite fuel, corporate fleets, business travel, energy purchased from utilities and employee commutes, they often don't include metrics on water, waste, procurement, toxics, supply chain impacts, or social impacts, among others, because those categories are not required under many reporting schemes.

2.1.1.1.2 Case Study – Corporate Carbon Footprint of a Shampoo Manufacturer (Scopes 1, 2, and 3)

The production of the electricity consumed by the plant is the major contributor to the scope 1 and 2 carbon footprint. The production of the raw material and ingredients, the distribution, use phase (consumption of warm water) and disposal of the products sold are responsible for 86% of the total carbon footprint (scopes 1, 2, and 3) of the company.

Scope 3 emissions can be categorized using an Economic Input–Output Life-Cycle Assessment model to identify upstream emission sources that are likely to contribute significantly to the company's total impact. A screening assessment using generic or literature-sourced data can help identify the "hot spots" in order to focus the data collection efforts. Experience demonstrates that the major portion of the total corporate carbon footprint can be accounted for by collecting full emissions information from only a limited number of suppliers.

2.1.1.2 Life-Cycle Assessment

A life-cycle assessment (LCA) is a method to assess environmental impacts associated with all the stages of a product's life from cradle to grave (i.e., from raw material extraction through materials processing, manufacture, distribution, use, and disposal or recycling).

The goal of an LCA is to model the full range of environmental effects assignable to products and services in order to improve processes, communicate impacts, and support policy and strategic decision making.

The procedures of LCA are part of the environmental management standards: in ISO 14040:2006 and 14044:2006.

2.1.1.2.1 The LCA Step by Step

According to the ISO 14040 [1] and 14044 [2] standards, an LCA is carried out in four distinct phases.

Goal and Scope

An LCA starts with an explicit statement of the goal and scope of the study, which sets out the context of the study and explains how and to whom the results are to be communicated. The goal and scope includes technical details that guide subsequent LCA process choices:

- the functional unit, which defines precisely what is being studied and quantifies the service delivered by the product system, providing a reference to which the inputs and outputs can be related;
- the system boundaries;
- any assumptions or limitations;
- the allocation methods used to partition the environmental load of a process when several products or functions share the same process; and
- the impact categories chosen.

Recommendations

A good practice in LCAs is to identify harmonized rules developed specifically per product category, by the applicable industry. These documents are called Product Category Rules (PCRs). In France, the working group of the environmental labeling of consumer products initiative is currently developing PCRs for personal care products.

To illustrate the notion of a functional unit, based on the discussions of this working group, a functional unit for the assessment of the impacts of shampoo in the USA could be:

- "1 washing of hair in the United States."

The reference flow could be:

- "1 dose of 8 grams of shampoo."

The scope of the LCA can be described using a process chart.

Life-Cycle Inventory

Life-Cycle Inventory (LCI) analysis involves creating an inventory of flows from and to the natural environment for the product system. Inventory flows include inputs of water, energy, and raw materials, and releases to air, land, and water.

The input and output data needed for the construction of the model are collected for all activities within the system boundary, including from the supply chain.

For product LCAs at either the generic (i.e., representative industry averages) or brand-specific level, that data is typically collected through survey questionnaires.

Primary and Secondary Data

The quality of the LCA directly depends on the level of detail of the data collection.

In order to reduce the time and to make LCA affordable, the simplification of the data collection is an important step. Each component of the LCA relies heavily on the quality and the relevance of the data. LCA data is usually categorized as primary or secondary data. Primary data is plant-specific or process-specific data that the LCA practitioner can directly access, and is involved in the data collection process. Secondary data is publicly available data that has not been collected specifically for the purpose of conducting the LCA, and for which the LCA practitioner has no input into the data collection process.

For example: to consider the impacts of the production of the body of a polyethylene terephthalate (PET) plastic jar, it is necessary to collect the exact quantity of plastic used (primary data). However, the impacts of the production of the PET plastic and of the injection molding process can be considered using industry average data (secondary data), and does not need to be specifically calculated for each facility.

Cosmetic products are very complex due to the diversity of the product ingredients, as well as the packaging.

In order to simplify the data collection, the section below describes the data that should be collected using measured information and the data for which industry average (LCI database modules) can be used:

Primary Data:

Composition of the product	Quantity of each ingredient (dry active matter) Quantity of water in composition of the formula
Composition of the primary packaging	Quantity and material type of the packaging including the cap % of recycled material in composition (post-consumer)
Production facility: Mixing and packaging of the product	Location of production facility Quantity and source of energy consumed Quantity and type of treatment of the waste
Transport of the product from the production facility to the store	Average distance and type of transport

Secondary Data:

Production of the packaging material (primary, secondary, and tertiary)	The production of the material can be considered using industry average data
Production of the ingredients	The production of the ingredients can be considered using industry average data
Use phase	Quantity of product staying in the packaging at the end of life, if any, (considered as waste)
Use phase for cleansing products (shampoo for example)	Quantity of water consumed and energy used to warm up the water
Production of the packaging	Production process of the packaging (molding by injection . . .)
End of life of the packaging	Recycling rates of the packaging material according to the national average considering the country of use
End of life of the ingredients	When the ingredients are released into the environment (cleansing products)
Secondary and tertiary packaging	Quantity of cardboard, pallets, and PET film used for the shipping
Transport of the ingredients and the packaging materials from the supplier to the manufacturing site	Distance and transport mode
Production process	Losses

The need to collect primary or secondary data may vary according to the product category as well as according to the objectives of the study.

2.1.1.2.2 Life-Cycle Impact Assessment (LCIA)

Inventory analysis is followed by impact assessment. This phase of LCA is aimed at evaluating the significance of potential environmental impacts based on the LCI flow results. Classical life-cycle impact assessment (LCIA) consists of the following mandatory elements:

- selection of impact categories, category indicators, and characterization models;
- the classification stage, where the inventory parameters are sorted and assigned to specific impact categories; and
- impact measurement, where the categorized LCI flows are characterized, using one of many possible LCIA methodologies, into common equivalence units that are then summed to provide an overall impact category total.

Characterization is the last compulsory stage according to ISO 14044:2006 [2]. However, depending on the goal and scope of the LCA, normalization, grouping, and weighting of the results may be conducted to facilitate the interpretation. In normalization, the results of the impact categories from the study are usually compared with the total impacts in the region of interest, the USA for example.

Grouping consists of sorting and possibly ranking the impact categories. During weighting, the different environmental impacts are weighted relative to each other so that they can then be summed to get a single number for the total environmental impact. ISO 14044:2006 [2] generally advises against weighting, stating that "weighting, shall not be used in LCA studies intended to be used in comparative assertions intended to be disclosed to the public." This advice is often ignored, resulting in comparisons that can reflect a high degree of subjectivity as a result of weighting.

Recommendation

The French environmental labeling initiative working group developing the Product Category Rules for Shampoos recommended three major indicators to be considered for the calculation of the environmental impacts of shampoos:

- Global Warming, expressed in g of CO_2 equivalent, according to the IPCC 2007 methodology.
- Water consumption, expressed in liters, corresponding to the net water consumption.
- Water toxicity*.

  *To date, there is no consensus about the methodology to apply for the calculation of the Water Toxicity but work is being done to address this issue.

In general, it is important to be very transparent with the choice of the methods and to be consistent with the method used in case of comparison of the performance of two products.

2.1.1.2.3 Interpretation

The results from the inventory analysis and impact assessment are summarized during the interpretation phase. The outcome of the interpretation phase is a set of conclusions and recommendations from the study. According to ISO 14040:2006 [1], the interpretation should include:

- identification of significant issues based on the results of the LCI and LCIA phases of an LCA;
- evaluation of the study considering completeness, sensitivity, and consistency checks; and
- conclusion, limitations, and recommendations.

A key purpose of performing life-cycle interpretation is to determine the level of confidence in the final results and to communicate them in a fair, complete, and accurate manner.

Interpreting the results of an LCA starts with understanding the accuracy of the results and ensuring they meet the goals of the study. This is accomplished by performing a sensitivity analysis on the data that contribute significantly to the impact categories and by assessing the completeness and consistency of the study. The last step consists of writing conclusions and giving recommendations based on a clear understanding of the limitations of the study.

2.1.1.3 Case study – LCA of Skin Care Products

2.1.1.3.1 LCA Tools and Uses

There are two basic types of LCA tools:

- dedicated software packages intended for practitioners; and
- tools with an LCA function operating in the background intended for people who want LCA-based results without having to actually develop the LCA data and impact measures.

In the former category, the principal tools are GaBi software, developed by PE International, and SimaPro, developed by PRé Consultants. Several other tools can also be well adapted for the development of an LCA of cosmetic products. The selection of an LCA software should consider various parameters, such as the level of expertise required to use the tool, the data available in the database, the cost and ease of use.

LCA is mostly used to support a business strategy or R&D, as input to product or process design, for educational purposes, and to provide support to the environmental communication strategy.

Companies are either undertaking LCA in house or commissioning studies. LCA can also be used for ISO Type III labels, called Environmental Product Declarations. This comprehensive third-party certified LCA-based label provides an increasingly important basis for assessing the relative environmental performance of products that compete on the open market.

2.1.2 ECODESIGN

2.1.2.1 Definition and Principles

Ecodesign is an approach to the design of a product with special consideration for the environmental impacts during the whole life cycle of the product.

Ecodesign aims at finding the best compromise between the various constraints related to the development of a product, such as cost, technical feasibility, regulatory requirements, and the client's expectations. Ecodesign principles are defined in the ISO 14062 standard [3].

Applying an ecodesign approach means including environmental considerations in product or service development policies. This process can create a competitive advantage from the significant added value of quantifiable and qualifiable product improvements. Ecodesign requires the participation of everyone involved in the development of new products, to improve the environmental performance and to create innovative solutions that can be highlighted for marketing purposes.

Ecodesign is a progressive and continuous improvement process. Companies that have experienced the deployment of quality processes will find many similarities in the deployment of Ecodesign.

A global and multi-criteria approach should be implemented in order to avoid transferring the impact from one environmental impact category to another; or from one life cycle stage to another.

2.1.2.2 Responsibilities and Tools

Ecodesign means design overall. Implementing Ecodesign requires adjusting the product development process by defining responsibilities and providing tools to the interested parties [3,4].

By starting with a pilot project, the company can gain experience, identify the interested parties to involve, test different tools, and understand the type of data required to measure the impacts, and so on. This experience can then be progressively extended to other product development projects.

Ecodesign requires the involvement of all the interested parties, including the marketing, R&D team, sustainability, logistics, and communication teams. Suppliers should also be involved in the process, especially when most of the impacts of products are in the supply chain.

Defining responsibilities and providing pragmatic and easy to use tools is key for the success of the implementation of Ecodesign. The first tools to support Ecodesign are the tools typically used by the design teams, such as technical benchmarks, market analysis, creativity sessions, value engineering, function analysis, and so on.

LCA is an important tool to support Ecodesign. However LCA should be used only when necessary, to support the decision making and confirm the environmental benefits of alternative design options. LCA should also be simplified in order to fit with the product development constraints in terms of time and cost. Alternative tools can be developed, based on the conclusions of LCA [4].

Environmental communication is a key tool to realize the benefits of the deployment of Ecodesign, therefore internal and external communication should be developed. According to the targeted market, companies can develop various types of communication strategies. The series of the ISO 14020 [5] standards provide companies with principles and guidance to develop fair, verifiable, and relevant environmental claims.

2.1.2.3 Case Study – Ecodesign of a Package

In this project, the company was working on the re-design of the packaging. Due to an increase in the cost of the raw material, the design team was looking for solutions to reduce the quantity of material used in the primary packaging without compromising the customers' perception of quality.

The LCA was conducted using a set of five impact categories covering the air, water, and soil. Findings corresponded to the comparison of the carbon footprint (global warming impact expressed in grams of CO_2 equivalent). The assessment illustrated that the new design allowed for a reduction in the carbon footprint of the product by 10%. This improvement represented a reduction of 100 grams of CO_2 eq. per product, which is equivalent to driving a Fiat Punto for 1 kilometer. Thus, the total amount of CO_2 saved for the annual production volume was significant.

The next step for this company is to consider the feasibility of replacing the glass with transparent plastic and to engage the suppliers in a continuous improvement process. The LCA will be used to design a sustainability index to rate and score the performance of the suppliers in order to source more sustainable raw material. The development of a strong partnership with the suppliers will allow for the collection of specific data with the range and accuracy required to improve the quality of the LCA, in order to prepare a consumer facing communication program.

2.1.3 GET READY FOR THE FUTURE

2.1.3.1 Environmental Labeling of Consumer Products

Policies recently introduced in France offer one possible model for organizations to follow. Since 1 July 2011, the Grenelle law on environmental protection has been implemented in France; it requires products to display information on their environmental impact, including the carbon emission of their whole life cycle (from material extraction, through manufacturing, packaging, stocking, and use, to disposal). This law was passed on 12 July 2010. A year later, more than 160 retailers and manufacturers in France (including some US brands) began to display information on the environmental impact of some of their products as part of a one-year experiment to find out how consumers respond to environmental labels. A working group is focusing on the development of Product Category Rules specific to health and beauty products, in order to harmonize and simplify LCA practices. Many challenges have been identified in the cosmetics industry, including the difficulty of accessing specific data to consider the impacts of the supply chain.

Being consumer facing seems to be an interesting goal and is an effective way of informing customers about the environmental impacts of selected products and the choices they can make. It empowers consumers to discriminate between products by identifying those that perform better. However, companies will have to first increase traceability within their supply chains to improve their understanding of the overall product impact, before being able to use LCA as a tool to differentiate their products in the market.

2.1.3.2 Programs to Involve Suppliers in a Continuous Improvement Process

Working conditions and environmental requirements in the global supply chain are becoming an increasingly sensitive topic. As described in the previous sections, most environmental impacts of products occur in the supply chain.

Suppliers are developing green claims to differentiate their products from the competition. So how can a retailer or a manufacturer validate a supplier's claims and choose a supplier when each product has been evaluated using a different method?

Many manufacturers, retailers, and third-parties are working on the development of methods to assess the social and environmental performance of suppliers.

Developing and implementing a sustainable suppliers program is a four step process.

2.1.3.2.1 Development of the Index

Before you start on the road to a green supply chain initiative, make sure you have all your assets on board including the upper management.

The first step consists of the development of a "Sustainable Supplier Index" tailored to your company's context. The design principles for the "Sustainable Suppliers Index" are the following.

The index must:

- Be compatible with existing industry standards.
- Have measureable impact:
  - Deliver business value to companies.
  - Drive quantifiable improvement.
  - Drive behavior change and promote continuous improvement.
  - Be easy to use and easy to understand for users (deliver results in a simple and useful format).
  - Engage the supply chain in measurement (including collection of primary data).
  - Be dynamic (adjusted over time to reflect new methods and data).
  - Enable "proactive" (pre-commercial) process/product design decisions and "reactive" end-product evaluation.
  - Weight supplier responses based on ability to drive impact (increase weighting for areas of greatest leverage).
- Be credible:
  - Employ a life-cycle-based approach that focuses on priority hotspots and uses fully transparent, widely agreed-upon, best available measurement methods.
  - Have a fully transparent scoring algorithm and enable third-party verification of results.
  - Use consistent, widely agreed upon, impact area weightings (not defined by your company but agreed by the industry when existing).
  - Focus on business-to-business decision-making first.

The index can be based on existing industry standards. It contains a set of questions to collect the relevant information in the supply chain. This questionnaire should be straightforward and focus on the major environmental issues.

A scoring system can be developed to rate the suppliers and measure improvements.

2.1.3.2.2 Pilot Project – Implementing the Index with Select Suppliers

A pilot project will allow for testing the index and identifying potential improvements. It also allows the company to begin to trace information within the supply chain. It is recommended that a project be initiated by identifying the major Tier 1 suppliers. However, it is acknowledged that the major impacts for the production of some ingredients or products may occur at a lower supplier level. Thus, in the medium term, the supply chain upstream of Tier 1 suppliers should be examined in order to collect information and score the suppliers, contributing significantly to the impacts of the products.

The company is usually in charge of collecting the information in the supply chain using a Sustainable Suppliers questionnaire developed in step 1. Avoid confusion from the start by bringing suppliers to the table and clearly explaining the business case behind the initiative and expectations for their participation.

Do not assume that your suppliers will all be novices; some may be further ahead of the curve than you are, giving you a chance to learn from them. It is important to have a record of where you are starting from and where you want to go. This means that if you want to help your suppliers reduce their energy consumption, water consumption, or waste production, you need to know what their emissions are so they can identify a solid strategy for improvement.

2.1.3.2.3 Engage Suppliers in a Continuous Improvement Process

Programs such as the Global Social Compliance Program (GSCP) provide tools and guidelines to implement environmental audits with suppliers that decide to receive external support. These audits involve measuring the current situation of the supplier and recommending action plans to progressively improve the energy, water, and material utilization efficiency.

Successful environmental initiatives will produce reciprocal benefits for both your company and suppliers. Your company will enjoy fewer risks and a more responsible supply chain. Suppliers will strengthen their business relationship with your company through better performance, improve the cost structure of their operations, and gain access to best practices and other information that can become a basis for innovation.

2.1.3.2.4 Progressively Extend the Scope of the Implementation of Your Index

Companies should target additional Tier 1 suppliers and should, when possible, try to involve Tier 2 suppliers.

It is worth noting that, for suppliers, when clients begin to ask about sustainability issues, it quickly becomes a top business priority. Undoubtedly, there are new conversations happening in companies across the globe that have been spurred by such indices. For many of these companies, it may be their first encounter with the concept of sustainability. After a short training process and a gap assessment, these companies are likely to start putting programs in place that are needed to score top marks on the index.

2.1.4 CONCLUSIONS

People involved in sustainability for more than 10 years have seen an evolution in the prioritization of sustainability actions within the industry. Companies started by addressing the low hanging fruit, monitoring and improving the energy and water consumption of their buildings, enabling them to achieve a respectable short-term return on investments. Today, most industries prioritize supply chain sustainability, including major brands in electronics, textile, food, and other industries.

The number of sustainability standards, labels, and product certifications has grown immensely, but what progress have we really made toward improving the life-cycle impacts of the products?

Monitoring supplier's performance will not be enough. The approach will have to include management system assessments and supplier development programs. These programs will include training and will provide suppliers with the technical expertise to setup individual goals and targets. Significant improvements of the supply chain can happen if suppliers receive a clear message from the market that sustainability performance will affect their business. Buyers will have to be trained to discuss sustainability with their suppliers during regular business meetings, supplier's summits, and product reviews in order challenge the vendors, while buyer sustainability purchasing performance is made part of their key performance indicator reviews. In other words, companies will need to integrate product and supplier sustainability considerations and information into procurement decisions.

In addition, some of the most significant impacts from a life-cycle perspective often occur in the raw material production and processing phases. The industry will have to face the challenges of working at multiple levels of the supply chain, from raw materials, to processing, manufacturing, and retailing.

As we look forward, it is clear that companies will have to face natural resource constraints and social pressure to act responsibly. Businesses will be held accountable for impacts that may be far removed from core operations, but still considered part of their company's supply chain sustainability performance. Metrics will be more and more standardized and data more readily available for improving sustainability impacts. If your company is looking for the next steps, these initial questions can be the starting point:

- How do you want your supply chain to look in 5, 10, 15, and 20 years' time?
- What are the goals and targets that you have to set to reach these objectives?

- What do you have to adjust in your supply chain sustainability strategy?
- Who are the key people who will be accountable to examine business as usual and identify areas for improvement?
- Which external allies, from investors to consultants to consortiums, will join you in addressing supply chain sustainability impacts?

ACKNOWLEDGEMENT

I would like to thank Michael Richardson for reviewing and commenting on this contribution.

PART 2: INNOVATING TO REDUCE THE ENVIRONMENTAL FOOTPRINT, THE L'ORÉAL EXAMPLE

Jean-Florent Campion, Rachel Barre, and Laurent Gilbert

2.2.1 INTRODUCTION

The L'Oréal Group has integrated the principles of sustainable development into all stages of a product's life cycle, from design to consumer use. For many years, this group has been committed to an eco-responsible approach with regard to the impacts of its activities, aiming at excellence in environmental performances. In 2011, clear progress in reducing the environmental footprint was achieved, focusing upon two strategic areas: the efficiency of resource use and technological innovative projects.

As part of its 10-year environmental strategy (2005–2015), the general management of L'Oréal assigned a precise target to its plants and distribution centers, worldwide: to reduce by 50%

- Greenhouse gas emissions (in absolute value).
- Water consumption per finished product unit.
- Waste production per finished product unit.

These environmental targets are ambitious considering L'Oréal aims to reach a billion new consumers in the coming years. The group realizes it needs to seize every opportunity for savings and rationalizations, as well as innovations in advanced technologies in line with the best environmental approaches.

2.2.2 PRODUCT ECO-DESIGN

Since 2005, L'Oréal Research has set the principles of green chemistry at the core of its eco-design approach. In 2011, the processes used to obtain Pro-Xylane (a derivative of a natural sugar extracted from beech wood) were further improved by reducing the amount of solvents used. Consequently its "E-factor" (the quantity of waste generated by the production of product) was reduced to 4.9, compared to an initial 13.0 and 6.6 in 2009.

Over the years, the group has increased the proportion of plant-based ingredients in its portfolio. In 2011, 55% of registered new ingredients were of vegetal origin, compared to about 40% in 2010. The same year, the amount of ingredients following the rules of green chemistry reached 45%, compared to 26% in 2010.

L'Oréal is dedicated to developing ingredients which have a low environmental impact. The group strives to develop ingredients of the lowest environmental impact, together with the best safety profile for consumers. To meet this goal, an integrated approach is used. Key objectives are adequate choice of ingredients, traceability of ingredients, evaluating their behavior vis a vis ecosystems, and their impacts on biodiversity. Since the environmental impact of cosmetic products is dictated by their ingredients, L'Oréal started to assess its entire raw material portfolio – based on PBT and vPvB criteria – in 2004. PBT denotes persistency, bioaccumulation, and toxicity, whereas vPvB describes very persistent and very bioaccumulative aspects. In 2008, 99% of its raw materials were evaluated.

L'Oréal has also developed an indicator for assessing the overall eco-toxicity of a given cosmetic formula by combining the individual contribution of each raw material.

The Sustainability Assessment Framework for ingredients. This tool allows manufacturing companies to assess the impact of ingredients on five main issues:

- Consumer and employee health
- Environmental protection
- Protecting biodiversity
- Fair trade
- Respect of human rights.

The framework is processed in close collaboration with suppliers to achieve ongoing progress covering the entire life cycle of substances, including environmental assessment. It comprises 25 assessment criteria based on the main international standards for sustainable development. This tool is now used with 120 substances and enables defining action plans with suppliers for minimizing the impacts of the ingredients upon all major issues.

2.2.2.1 Predictive Evaluation and Green Chemistry

The concept is simple: to generally reduce, as much as possible, both the amount of substances used during tests and the substrate required to conduct them. For example, when testing a hair color product, the evaluation teams used wigs of several hundred grams of hair or locks of a few grams for decades. Reducing the consumption of both hair and chemical products led to a gradual decline in the quantities of hair used: down to mini-locks, then to small test tubes samples, resulting in the minimum use of the hair powder. Nowadays, only 10 milligrams of hair and just a few milligrams of chemical products are used for testing hair color, hair care products, and make-up formulas (like mascaras). Initially, this systematic scaling down was started by the Department of L'Oréal Life Sciences, which has long been conducting routine tests at micro-molar levels or even lower. L'Oréal approach, therefore, involved extrapolating these concepts towards physical testing that is closer to the actual usage levels of cosmetics products.

Respecting the fundamental principles of green chemistry has become crucial in developing new ingredients. In 2010, L'Oréal used approximately 40% (in volume) of ingredients from renewable vegetal-based ingredients in its finished products. These renewable ingredients are in all types of product applications – from skin and hair care products to colors and make-up. In order to efficiently select our new processes and ingredients and to improve our existing processes, the level of green chemistry [6] has to be imperatively determined.

L'Oréal set up the following list of "green indicators" for defining the advanced eco-design of its future ingredients:

- "Atom economy" calculation [7].
- The E-Factor, which will constantly be lowered during scaling-up and industrialization [8].
- The level of renewable carbon in the final ingredient.
- Environmental risk assessment of the final compound is carried out according to the EU guidelines on requirements in the PBT Assessment. Obviously, substances with PBT or vPvB profile will not be developed.

These green indicators are also used for ingredients already used in formulations to re-evaluate their environmental impact. Six ingredients of the highest volume of use have an E-factor ranging 0.36–3.3 indicating levels of waste reduction.

The task of reducing the environmental impact of ingredients is by essence a continuous one, fully integrated into the objectives of green chemistry.

In 2009, the improvement of two industrial polymer processes led to the lowering of their E-Factors by 51 and 35%, respectively. This work enabled L'Oréal to reduce waste by several dozen of tonnes per year, just by a decrease in solvent consumption.

To ensure energy efficiency, processing that incurs high energy costs is avoided, such as:

- Temperature < -15 °C or > 150 °C.
- Duration > 10 hours.

All these indicators clearly contribute to the advanced selection of ingredients and processes with the lowest environmental impact, enabling them a possible access to market.

2.2.2.2 Biodegradability of Formulas

To constantly minimize the environmental impact of ingredients, L'Oréal research teams look at improving the biodegradability of formulas. This property is evaluated by sorting out the total percentages of biodegradable ingredients. Once calculations are made, strict internal criteria are applied to give a "biodegradable" status to the formula. For example, in 2011 the group launched the 97% biodegradable Ultra Doux Almond and Lotus Flower shampoo and conditioner, and the 94% biodegradable Fructis Pure Brilliance shampoo and conditioner. Many products have been eco-designed, especially rinse-off products; they include The Body Shop Rainforest shampoos and conditioners, Dop Nature shampoos and conditioners, and Biotherm "Ecume de douche" shower gel.

2.2.2.3 Eco-Design of Packaging

Since 2007, L'Oréal has been committed to a strict policy of using sustainable materials for its paper and cardboard packaging. Forest Stewardship Council (FSC)-certified materials are produced via responsible forestry operations that guarantee the forest's sustainable management, commitment to the employment of local and regional workforces, and the enforcement of stringent safety standards.

L'Oréal also looks at reducing the weight and volume of its packaging and to lowering its impact on the environment. So, every time new packaging is considered, a simplified comparative life-cycle analysis is conducted to select a design that will help to reduce these effects. The study takes into account all the stages of the packaging life cycle, based on eight impact indicators (including CO_2, water, eutrophization, waste, etc.).

For several years, the Garnier brand has been involved in an eco-design initiative for its products and packaging consisting of successive stages. The iconic 50 ml skin care jar initially used 40% recycled glass, reducing the energy consumed to produce the jars. For its latest renewal in 2011, Garnier has a lighter plastic design which is even more attractive, whilst reducing the packaging impact on the environment by 30%. This assessment was carried out, according to the life-cycle analysis, by an independent market research firm and confirmed by a jury of experts, in accordance with the ISO 14040 and 14044 standards.

The Biotherm brand saves several tonnes of packaging by using the eco-design approach. In 2011, 80% of the 50 ml glass jars used for skin care cream contained recycled glass (up to 40%). All (100%) of the cardboard packaging was certified (70% FSC and 30% PEFC, with a target of 100% FSC). The elimination of instruction leaflets, now printed on the inside of the box, saved 24 tonnes of paper in 2011 (the expected saving is 30 tonnes in 2012). The plastic PE bottles were also revamped: a 6 g reduction in the 400 ml bottle, from 40 to 34 g, and a 2.5 g reduction for the 200 ml bottle, from 25 to 22.5 g. Combined, these reductions have resulted in 10 tonnes of plastic saving in 2011, and an expected saving of 26 tonnes in 2012. Biotherm, a community spirit brand, encourages consumers to recycle by using twist-off pumps; it also provides helpful information on the packaging.

2.2.3 RESPONSIBLE SOURCING AND BIODIVERSITY PRESERVATION

Respecting, protecting, and valuing biodiversity are the cornerstones of L'Oréal's worldwide policy of sustainable sourcing. In fact, L'Oréal was among the first companies worldwide to commit to operating in full respect of the principles of the Convention on Biological Diversity.

Well before the Nagoya Protocol came into effect in 2010, L'Oréal Research approached the issues of Access and Benefit Sharing very seriously. The group believes that the responsible use of biodiversity offers a vast potential for innovation (e.g. novel ingredients, bio-inspired processes), which is highly valued by consumers, especially those located in biodiversity-rich countries; it is also a great lever for social inclusion in supply chains.

Since 2005, L'Oréal has been increasing the use of renewable vegetal-based ingredients in its products. In 2011, 55% of new ingredients registered were of plant origin, compared to 40% in 2010. As well as increasing the share of renewable ingredients, L'Oréal Research has strived to secure and build supply chains to meet the needs of sustainable development.

Thus, the development of any innovative product arising from biodiversity is preceded by a thorough investigation on the local sourcing in the country from which the products originate.

Assessing the impact on biodiversity of ingredients of vegetal origin requires the use of a number of operational tools to identify any ecological or social challenges associated with any given plant. Among them are:

- Knowledge and Risk indicators (the "plant data sheets"). These indicate the ecological status of a resource, the degree of benefit to the community, any change in land use resulting from production of the raw material, and so on.

- A multi-criteria matrix (the "Raw Material Sustainability Assessment Framework"), for assessing suppliers' practices along supply chains.
- A set of external guarantee schemes and expert third parties, to lead specific actions on the field; these include FSC, RSPO, FLOCert, BioSolidaire, UEBT, and so on.

In 2008, 100% of renewable ingredients were evaluated on biodiversity criteria. All new raw materials now come from supply channels that offer reliable guarantees of traceability and sustainability.

In order to guarantee the recognition of traditional uses and know-how within the management of all intellectual property aspects, L'Oréal carried out a systematic and thorough review of its patents in 2009. This was done to make sure they were respecting traditional know-how and that knowledge was made accessible to the local population. The same criteria were applied to the patents of suppliers.

This enabled L'Oréal to formalize and document best practices, to help improve the livelihoods of local people involved in extracting the natural raw materials, while conserving biodiversity.

2.2.3.1 Case Study: Palm Oil

Although L'Oréal is a rather low volume purchaser of palm oil – direct annual consumption did not exceed 1000 tonnes in 2011 – the issue of sustainable sourcing has nevertheless been taken very seriously.

L'Oréal has been a member of the Roundtable on Sustainable Palm Oil (RSPO) since 2007. Since 2010, it has sourced 100% of its palm oil from Certified Sustainable Palm Oil sources (CSPO), following the demanding system of traceability offered by RSPO (Segregated Model). L'Oréal has always supported the moratorium on illegal deforestation in Indonesia and Malaysia; it now plans to improve the traceability of its palm oil derivatives. This category of compounds are a very challenging issue for the personal care sector because of the difficulty in tracing the complex supply chain, and in understanding how the derivatives were processed from their initial source. In the Coalition of Sustainable Palm Oil Members, L'Oréal has come up with a reasonable approach on how to measure indirect palm oil consumption. For this category of compounds (mainly used in foaming hair care applications) L'Oréal believes the RSPO Book and Claim model constitutes a workable interim option during the transition period to physical use of certified oil. This is why L'Oréal is committed to source 100% of derivatives from certified sustainable sources by 2012.

For the second time in 2011, L'Oréal's palm oil sourcing strategy received recognition by the WWF with a score of 9/9 on their "Palm Oil Buyer Scorecard" rating scale.

2.2.3.2 Vercors Botanical Garden

The access to a botanical garden located in the French region of Vercors enables L'Oréal researchers to study more than 350 plant species.

The move towards organic and natural products adopted by L'Oréal Research has led to numerous formulas labeled "organic Ecocert." The ingredients portfolio has also been extended to include new organic materials. Around 500 ingredients in the group portfolio are currently in line with Ecocert organic standards.

The activity of the L'Oréal laboratory dedicated to natural and organic cosmetics is supported by an experimental farm and botanic garden located in the Vercors region of France. These facilities favor the development of products in line with organic farming, enabling more than 350 plant species to be studied. The laboratory also plays an active role in protecting biodiversity by diversifying the portfolio of plants used in its products, combating intensive monoculture, and promoting the cultivation of marginalized species. For example, catnip has recently been put back into cultivation for its anti-bacterial properties. In 2009, Sanoflore was the first French organic cosmetic brand awarded the Bio Solidaire certification, which recognizes supply chains that prioritize the sourcing of local ingredients.

2.2.3.3 The Solidarity Sourcing Program

Another emblematic approach, the Solidarity Sourcing program, shows how sustainable sourcing of biodiversity is a powerful lever for social inclusion when it impacts the renewable ingredients purchase category.

Inspired by The Body Shop's sustainable development and fair trade values, L'Oréal initiated the Solidarity Sourcing program in 2010. This purchasing program reaches out to fair trade producers, companies working for the inclusion of disabled people, companies promoting social integration, or very small businesses looking to advance social inclusion by helping economically vulnerable people to find work and income, in all countries where the group operates.

The flagship project of the program is the sustainable procurement of shea by a socially responsible subsidiary in Burkina Faso. Shea is an essential raw material for L'Oréal; the group's purchasing volumes keep rising. One characteristic of the raw material is that it is only present in sub-Saharan Africa, a region strongly impacted by climatic changes and where people have very low incomes.

The group has worked on structuring the subsidiary with two major suppliers with the objective to purchase the nuts directly from the harvesting communities. As well as cutting out the middleman, the move would tailor specific producer organizations and supply programs, and develop a unit for in situ processing. At the end of the dry season (April–May) when food reserves are almost exhausted, the harvesting women benefit from pre-financing of their production; higher purchase prices are given to tide them over.

In 2012, shea was in the top 10 list of L'Oréal plant-based ingredients. From the very first year, more than 50% of group purchases originated from the "Solidarity Sourcing" program. In total, 13 000 women should directly benefit from the L'Oréal "Solidarity Sourcing" program. By 2013, 100% of procurements will be socially responsible.

2.2.4 RESPONSIBLE PRODUCTION

2.2.4.1 Energy and Greenhouse Gases in Manufacturing, Water Use, and Waste Management

The cosmetics industry has a relatively low energy demand compared to other industries. L'Oréal is therefore exempt from European regulations on carbon emission quotas. However, the group is committed to using energy efficiently and reducing dependency on fossil fuels. As far as possible, L'Oréal purchases natural gas rather than fuel oil (which has higher carbon content), and continues to develop its strategy on renewable energy.

The group values each initiative taken by its local sites across the world. Each saving is important in reducing the global carbon footprint. In total, greenhouse gas emissions have been reduced by 29.8% between 2005 and 2011.

Most of the water used in L'Oréal factories relates to needs in cleaning production equipment and packaging lines, and for maintaining strict hygiene standards. This represents approximately 40% of all water consumption in the group's industrial sites.

To meet its targets, L'Oréal strives to reduce the amount of water used in all areas (cleaning, sanitary, cooling, etc.). A new method called OptiCIP has been developed, enabling the factoring in of specificities of a given industrial site – such as equipment, types of products – and further applying the most efficient and specific cleaning processes.

In 2010 and 2011, several factories launched pilot projects, with initial findings showing significant reductions of 50% and more on certain equipments. Implementation programs are now in place, aiming at applying the OptiCIP method to all manufacturing equipment.

In total, water consumption per finished product decreased by 22.6% between 2005 and 2011.

For many years, L'Oréal has followed a strong policy of waste management. This goes beyond regulatory compliance, consisting of waste prevention, recycling, and re-use, as well as energy recovery, and avoiding disposing waste to landfill.

In 2011, in line with the goals set of reducing waste per finished product by 50% (2005–2015), the group launched new initiatives that go further in reducing waste at source while reducing the overall environmental footprint.

At present, 95.7% of the waste is re-used, recycled, or recovered for the production of energy. More than 50% of the sites do not dispose waste to landfill. Transportable waste generated per finished product decreased by 24.2% between 2005 and 2011 (excluding returnable packaging).

Waste is defined as any materials, other than products, leaving L'Oréal sites. The group distinguishes different types of waste in accordance with European regulations.

These include transportable wastes (including cardboard, paper, plastic, dirty packaging, pallets, and manufacturing residues), returnable packaging for transport that circulates between suppliers and sites, and exceptional waste such as construction debris.

2.2.4.2 Certification

At present, 88% of L'Oréal's plants are ISO 9001 certified (2000 version) or possess FDA quality certification. Of the total, 85% are OHSAS 18 000 or VPP safety certified and 90% have the ISO 14000 environment certification.

2.2.4.3 Transport and New Carbon Reduction Initiatives

L'Oréal optimizes its global production on a regional basis, placing operation teams as close as possible to the markets they serve. This allows each production site to be more responsive and more efficient in terms of logistics and transport.

While greenhouse gas emissions in the cosmetics industry are relatively limited, transport still contributes to L'Oréal's overall carbon footprint [9].

L'Oréal's logistics network moves over 5.78 billion products a year from factories, to distribution centers, to the customers all over the world. Road represents 58% of the mode of transport used by L'Oréal. As far as possible, the logistics teams use greener forms of transport that enable the group to meet required costs and deadlines. In 2011, L'Oréal placed priority on other methods and different modes of transport with lower greenhouse gas emissions [7].

2.2.4.4 Environmental Performance of Factories and Distribution Centers

In 2005, L'Oréal committed itself to a 50% reduction in CO_2 emissions, water consumption, and waste production by 2015. Thanks to the operations teams, CO_2 emissions, water consumption, and waste production per finished product have already been reduced by 29.8, 22.6, and 24.2%, respectively. These results are also due to sustainable innovation: from plants to distribution centers, all our sites have rallied round to find creative, effective solutions.

2.2.5 REFERENCE ACTIONS – SOME EXAMPLES OF KEY ACHIEVEMENTS

2.2.5.1 Shampoos Ultra-Kind to the Environment

Designing an environmentally friendly product implies factoring in its entire life cycle and applying environmental excellence criteria at every stage. This was done by L'Oréal with the Garnier Ultra Doux range (see Figures 2.2.1, and 2.2.2), which was completely overhauled in 2011; it was part of an environmental labeling experiment conducted jointly by Garnier and Carrefour.

L'Oréal's eco-design approach, now the focus of all stages of development, is illustrated by the choice of ingredients, automated trials on minute quantities to limit waste, the enhancement of "co-products," manufacturing processes that reduce energy and toxic solvent use, and recyclable packaging.

Among others, one task of the L'Oréal Research teams deals with improving the biodegradability of formulas. Several initiatives were launched in 2011, namely in Garnier's Ultra Doux and Fructis ranges, to constantly minimize the environmental impact of ingredients. This property is assessed by working out the total percentages of biodegradable ingredients. Once the calculations are made, strict internal criteria are applied to award "biodegradable status" to the formula. The 97%

Figure 2.2.1 Biodegradable Ultra Doux shampoo with its refill system.
Source: Reproduced with permission from © L'Oréal.

Figure 2.2.2 Biodegradable cosmetic products.
Source: Reproduced with permission from © L'Oréal.

biodegradable Ultra Doux Almond and Lotus Flower shampoo and conditioner, and the 94% biodegradable Fructis Pure Brilliance shampoo and conditioner were launched in 2011.

Within this ambitious framework, L'Oréal decided to review the Garnier Ultra Doux range using this environmental excellence criteria as a yardstick. The challenge for L'Oréal research was to develop a new formula while reducing its impact on the environment by checking the ingredient quantities. They also had to factor in and respect biodiversity and limit waste production, while reducing the quantity of packaging, encouraging the use of refills and recycled materials, and ensuring that communication on the products was in line with the various criteria.

To fulfill this mission, the teams used the L'Oréal eco-design charter which pinpoints five areas for vigilance throughout the product's life cycle: maintaining people's health, respecting the environment, protecting biodiversity, promoting fair trade, and managing the social impact of innovations.

To develop this eco-designed product, the challenge was to ensure that the cosmetic properties and quality of use were comparable to other classical shampoos. The criteria for creating an eco-designed formula are first based on the choice of ingredients. Naturalness, biodegradability, absence of impact on biodiversity, and low ecotoxicity are sought. The final formula must be very largely biodegradable ($\geq$95%) and not ecotoxic to aquatic organisms.

2.2.5.2 Environmental Labeling: A Joint Initiative between Garnier and Carrefour

In July 2011, Garnier joined forces with Carrefour for a year-long national experiment on environmental labeling launched by France's Ministry of Ecology and Sustainable Development. It involved giving consumers a reliable, understandable indication of the environmental impact of the complete life cycle of shampoos and their packaging: greenhouse gas emissions, water consumption, and water pollution (see Figure 2.2.3). In a large number of France's Carrefour supermarkets, "spotters" were placed by Ultra Doux shampoos, inviting consumers to visit web sites for further information.

Through jointly conducted experiments, L'Oréal and the Carrefour Group wanted to test labeling based on modeling the environmental impact of products throughout their life cycle. This test should assess the relevance of this type of information by first and foremost examining:

- consumer understanding and adherence,
- the possible impact on consumer purchasing behavior,
- the reliability of the information given,
- the compatibility and consistency of information with existing labels,
- the cost of roll-out from an environmental benefit viewpoint.

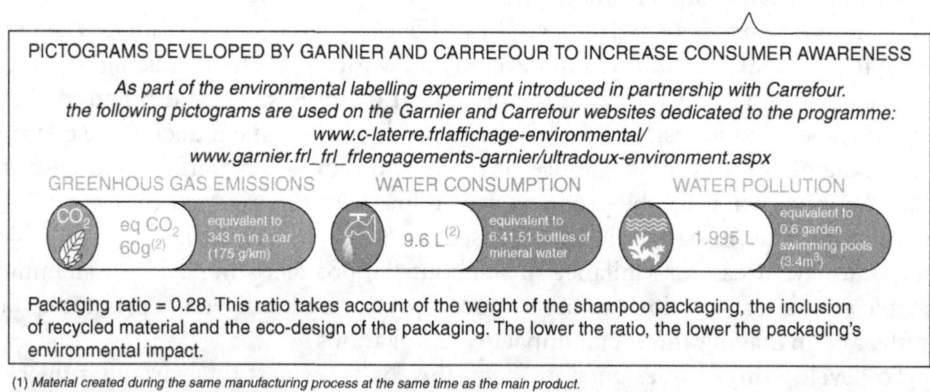

Figure 2.2.3 Ultra Doux shampoo environmental footprint.
Source: Reproduced with permission from © L'Oréal.

This voluntary experiment was conducted under the auspices of the French Ministry of the Environment. A report on the experiment will be submitted to the French Parliament. Through this testing, Carrefour and L'Oréal will contribute to public debate.

This experiment places consumers at the heart of the information campaign. It helps verify consumer adhesion: using an online survey, consumers can provide their own opinion on the labeling. L'Oréal and Carrefour wish to find methods of providing information that are most appropriate to meet consumer needs. This experiment provides a deeper understanding of what is needed to establish reliable, comparable environmental information on products: the international harmonization of calculation methods and the accessibility of databases are crucial for providing consistent information and preventing unfair competition between companies.

Finally, this collaborative effort produced representative feedback by combining a retailer with a large industrial group on a project: thanks to the experience of a group like L'Oréal, a leader in its sector, coupled with the energy of Carrefour as innovative supplier, both were able to get the most out of this test.

2.2.5.3 Responsible Management of Paper and Cardboard

For five years, L'Oréal has been implementing a stringent forestry certification policy for its packaging. It declares that over 90% of its cardboard boxes are certified, more than half by the Forestry Stewardship Council (FSC). At the same time, the group took actions aimed at reducing the weight and volume of its packaging and lessening its impact on the environment.

Central to innovation within the group, packaging plays a key role in L'Oréal's ambitious environmental approach.

From one end of the life cycle to the other, it is continuing its journey of transformation on the path to eco-friendly design. Since 2007, L'Oréal has been committed to a strict policy of forestry certification for its paper and cardboard packaging through the active promotion of FSC certification.

Previously only applied to wood, this international standard is now extended to cardboard and paper. Produced via responsible forestry operations that guarantee the forest's sustainable management, FSC-certified materials are committed to the employment of local and regional workforces and the enforcement of stringent safety standards.

FSC is a socio-environmental label which guarantees the preservation of the forest and combats deviations from labor law, such as the infringement of civil or human rights or the use of an illegal workforce. It is a guarantee of ethical practices and quality. Today, more than 90% of cardboard boxes are certified, over half by the FSC. The same approach has now been adopted by printers as well; 92% of their sites are now FSC-certified.

Every time an item of packaging is updated, a simplified comparative life-cycle analysis is performed to reduce these effects. This study takes into account all the stages of the packaging's life cycle and is based on eight impact indicators, including CO_2, water, eutrophication, and waste.

2.2.5.4 Sustainable Sourcing of Argan Oil

The program for the sustainable sourcing of argan oil was set up with the objective of ensuring a fair return to local communities, and better traceability and quality of ingredients. The program also sought to ensure the extraction of this oil was harmless to local biodiversity.

The argan tree (*Argania spinosa (L.) Skeels*) is endemic to southern Morocco, largely found on the plain bordering the Sahara desert (Souss region). It is the second largest forest eco-system in Morocco, covering about 2% of the country's superficy and providing habitat for a wide variety of wildlife species. The forest acts as a natural barrier against the advance of the desert and is paramount in preventing erosion and protecting water resources. The argan forest also plays a very strong socio-economic role, as it represents a major source of income for about 6% of the rural population and up to 90% of the economy in areas of native argan stands. Argan oil is extracted locally from the kernels of the fruit, which are handpicked by women of local communities.

The argan forest eco-system, threatened by over-exploitation, soil erosion, desertification, and changes in agricultural practices, is experiencing an estimated annual decline of 1%. As a result, UNESCO classified the argan forest as a biosphere reserve in 1998.

In 2007, L'Oréal signed a three-year pioneering tripartite partnership with its supplier BASF and a non-governmental organization, Yamana, in Morocco, to foster corporate social responsibility in the argan supply chain to procure argan oil.

In 2010, argan oil was added to hair care products, whereas kernels are used in scrubs, and argan cake and leaves extract are used in skin care products.

As well as improving quality of life for female workers and their families, the program is contributing to the overall economic and social development of the area. In 2011, Taraganine cooperatives became the first network of women's cooperatives for argan in Morocco. Today, there are more than 200 women working in 6 production cooperatives, 1 extraction and oil facility and 15 preparation cooperatives. Working conditions have been improved for 300 women thanks to an increased income, access to healthcare, literacy, and better hygiene and safety at work.

Consequently, women in this rural area have increased their financial autonomy, reinforced their decision-making power, and improved their social status. The program has also increased community awareness of the value of the argan eco-system and provided incentives for its conservation. Today, argan is present as an ingredient in all the group's product categories.

Figure 2.2.4 Phytoremediation garden.
Source: Reproduced with permission from © L'Oréal.

2.2.5.5 Phyto-Remediation Garden at Chimex

Another innovation came from this French subsidiary of L'Oréal, Chimex, specializing in the production of cosmetic ingredients. It had already come to symbolize the group's good environmental performance when use of water is at stake. This L'Oréal subsidiary synthesizes cosmetic ingredients using chemistry and biotechnologies and is a perfect example of environmental innovation and performance in the use of water. Six years of studies were needed to combine a physicochemical system with treatment by phyto-remediation (see Figure 2.2.4). Today, the "100% clean water" project makes it possible to treat the wastewater from a Chimex factory so effectively that it can be returned back to the natural environment. A first and important action for the chemical industry.

2.2.6 CONCLUSION

This contribution aimed at illustrating the commitments of an industrial group such as L'Oréal, towards what is generally depicted as "sustainable development." These commitments first arose from a clearly and strongly expressed policy to integrate all its criteria within its various activities. As expressed in the text, it is clear that such a long-term policy brought – globally and locally – significant improvements in water, waste, energy savings, and so on without any compromise on the innovation, the safety, and the quality of the products that L'Oréal offers to consumers worldwide.

Global decisions, local and specific adaptations at all levels are key. They have led L'Oréal to make significant improvements in social, economic, and environmental policies. Although the paths of these advances must remain continuous and open to new technologies, they now represent and fulfill, at the planetary scale, the needs of public interest, respecting economic growth, social and cultural wealth, integrating all environmental issues and preserving natural resources.

ACKNOWLEDGEMENTS

We would like to thank all L'Oréal participants for their contribution to sustainable development at all levels. Many other initiatives are taking place at every level and in all places where L'Oréal employees are daily contributing to the business of beauty. This contribution cannot obviously present all these actions, but it is a perfect opportunity to thank all L'Oréal employees for their formidable engagement and contribution to a sustainable future.

REFERENCES

[1] ISO (2006) 14040. Environmental management – Life cycle assessment – Principles and framework. International Organisation for Standardisation (ISO), Geneve.
[2] ISO (2006) 14044. Environmental management – Life cycle assessment – Requirements and guidelines. International Organisation for Standardisation (ISO), Geneve.
[3] ISO (2002) 14062. Environmental management – Integrating environmental aspects into product design and development. International Organisation for Standardisation (ISO), Geneve.
[4] Lacoste, R., Robiolle, M., and Vital, X. (2011) *Ecodesign of Electronic Devices*, DUNOD, Paris, p. 224 (French)
[5] ISO (2006) 14020. Environmental labels and declarations – General principles. International Organisation for Standardisation (ISO), Geneve.
[6] Philippe, M., Didillon, B. and Gilbert, L. (2012) Industrial commitment to green and sustainable chemistry: using renewable materials & developing eco-friendly processes and ingredients in cosmetics. *Green Chemistry*, **14**, 952.
[7] Trost, B.M. (1991) The atom economy – a search for synthetic efficiency. *Science*, **254**, 1471–1477.
[8] Sheldon, R.A. (1994) Consider the environmental quotient. *ChemTech*, **24**, 38–47.
[9] For more details, see page 24 of the L'OREAL Sustainable Development Report 2011.

3

The Social Impact of a Cosmetics Company

Bas Schneiders and Frédéric Anklin

3.1 THE RELATIONSHIP BETWEEN COSMETICS AND SUSTAINABILITY

Sustainability is presently seen as a key driver in the cosmetics industry and in the beauty sector overall.

Parallel to the rising level of prosperity worldwide the economic importance of the beauty business has increased considerably in the past decades. As such, its social and ecological impacts have also grown significantly.

In the cosmetics industry, social responsibility plays a special role, as cosmetics are at their core about human well-being and the desire for attractiveness and beauty.

On a much deeper, more symbolic, level there is a functional analogy between the beauty industry and sustainability.

Using cosmetics is usually a question of caring about yourself, attractiveness, looking good and looking younger, slowing down the process of decline. Working in a sustainable way plays a similar role, caring for the beauty of the planet and the human being.

Let's not fool ourselves, the signs are not good: loss of biodiversity and soil fertility, pollution and the overexploitation of natural resources, hunger and suffering,

Sustainability: How the Cosmetics Industry is Greening up, First Edition. Edited by Amarjit Sahota.

wars and violence. A world where more and more people are excluded and social inequality grows, to name only but a few challenges that humanity is facing.

However, sustainability offers a path which may allow us to achieve both, creating a more social and ecological balance against all odds. They key question is: what can the beauty sector contribute to make this world a better place?

3.2 THE GROWING SIGNIFICANCE OF SUSTAINABILITY

3.2.1 Current Situation

Today's societal and ecological conditions are more ambivalent then ever before.

On the one hand we are living in an almost totally globalized world of enormous wealth, that has seen prices of traditional luxury items and seasonal goods tumble and has made them available to the wider masses:

- With growing access to resources and increased global reach, people have become more conscious of associated environmental and health issues.
- The awareness for health and sustainability has evolved and the motivation to leverage conscious consumption for a better world is spreading from special interest groups to the mainstream.
- A perceived lack of meaning spurs the renewed interest in spirituality and the search for purpose. There is a longing for gentler medicine, harmony and nature.

On the other hand, it is becoming increasingly clear that the current lack of meaning is systemic and cannot be compensated for through consumption.

Socially, the public has lost trust in institutions and the great authorities:

- With economic divide on the rise, even in rich countries, exclusion and the opportunities for people to participate in societal affluence is becoming one of the biggest challenges of the twenty-first century.
- Through specialization and globalization the world is becoming more and more complex and confusing. The direct link between people and their environment, their actions and their consequences is disappearing. With it the feelings of self-determination and meaning, and often also responsibility are lost.
- Ecologically most signals are set to red. Humans are living beyond their means. The effects of climate change, rapid loss of biodiversity and soil fertility are only the tip of the iceberg.

- The economic drive towards "bigger, better, faster" is subject to intensifying criticism. It is becoming increasingly apparent that growth is no end in itself and growth cannot replace the need for meaning and happiness in life.

The following corresponding societal trends and drivers further increase the relevance of this topic [1]:

- The increasing importance of ethical consumption, for example, the way the lifestyle of health and sustainability is moving into the mainstream. This is particularly important for products that are used on the skin or internally.
- CSR (CS, CR, CC etc.) as a competitive advantage: these offer additional ways to provide added value to brands and companies.
- Pressure from the media and NGOs: the sustainability topic is prone to scandals and makes good headlines. Mistakes can lead to legal consequences.
- Changes in the environment and logistics: As people become more aware that resources are finite it is important to optimize the "sustainable footprint."
- Laws and regulations: important factors here include stricter environmental laws, stricter reporting requirements and minimizing risk.

3.2.2 Solution Strategies

In the face of the above circumstances, a wide variety of potential solutions are controversially discussed. The spectrum ranges:

- from highly regulatory to economically very liberal approaches that assume the economy can best contribute if left to self-regulate,
- from efficiency focused approaches to a complete life-cycle approach (cradle-to-cradle model),
- from post-growth economy strategies, some of which see economic decline as the path to sustainability, while others assume that only economic growth can generate the capital needed for a sustainable transformation of society,
- from technocratic management approaches to a more cultural approach, the emphasis of living well and fairly.

Regardless of the industry, the key to all potential solutions is to fully integrate sustainability into the core business, instead of treating it as a nice to have added value. This strategy unites experts, internal and external stakeholders, as well as consumers.

The complexity of sustainability forces humans into moral oversimplification. In relation to business this boils down to a single basic question that applies to any company: what does the company do to make this world a better place?

For a professional sustainability discourse, however, keeping the subject broad and looking at everything as an integrated and interconnected whole is crucial to develop comprehensive solutions.

The longing for simplicity is especially problematic for the social dimension of sustainability. For example, it often leads to a reduction of the CSR fields of action to HR relations, ethical supply chain and corporate philanthropy [1].

This is, however, too narrow a focus.

Furthermore it is important to understand the social functions performed by each particular industry and the potential and implications they hold for sustainability.

3.3 SUSTAINABILITY AS A SOCIAL CHALLENGE FOR COSMETICS COMPANIES

3.3.1 Social Footprinting

In light of the complex requirements for a positive sustainability performance laid out above, the challenge is to understand all relevant social impacts of a company. Those that are negative should not only be avoided, but where possible transformed into positive and meaningful impulses.

Similarly to the ecological footprint, a social footprint can help to illustrate a company's impacts, as well as its social costs and their effects (rejection of the input-orientation).

At the time of writing (January 2012) there is no entry on Wikipedia for "social footprint," which suggests the concept is novel. The innovative approach is a reason for hope.

The downside of its novelty is that established tools for measuring or monitoring social footprints are not yet available.

For now, related fields offer concepts and methodical approaches that could be adopted, for example from calculations for the social return on investment (SROI) or output and outcome evaluations.

Additionally GRIs social indicators may provide further orientation.

3.3.2 Critical Areas with Social Impacts

3.3.2.1 Ecological Footprint

It might seem odd to encounter the term "ecological footprint" under the social footprint heading. However, given the substantial amounts of materials and energy captured in the ecological footprint, it is also highly relevant in a social context. It shows to what extent a cosmetics company obstructs or promotes the fair use of global resources.

3.3.2.2 *Product Range and Pricing*

The call for integration of sustainability into the core business does not only pertain to the internal company perspective (HR, sourcing, production conditions), but also the external one (market, consumers, and their motivation) and more far-reaching questions:

- Which products are produced and marketed – and for what price?
- What is their availability?
- Who is – for example, through pricing – excluded from the use of these products (violation of the inclusion principle)?

3.3.2.3 *Relationships with Partners: Ethical Sourcing and Fair Trade*

Fairness in all relationships is a key factor of a social footprint, for which sensitivity has greatly increased in recent years.

One aspect is fairness with regards to business partners or external stakeholders in ever more complex supply chains. It also includes aspects like fairness in the choice of location for production, for example based on savings in taxes or wages. Due to the growing complexity of supply chains it is one of the most difficult areas of social sustainability.

One way for cosmetics companies to address social challenges within the supply chain is by building fair trade partnerships. Many companies have been working in such partnerships for decades, as it allows them to secure their supply of natural and organic ingredients whilst working in a socially responsible manner and protecting the environment.

3.3.2.4 *Information and Communication*

On the one hand it holds true that "Actions speak louder than words" and communicating sustainability without substance is greenwashing.

On the other hand, sustainability can be a differentiator and using good sustainability performance in communication can motivate employees and external stakeholders: "Do good and talk about it!"

Companies for whom sustainability is a founding principle and has been an integral part of their core business from day one run the risk of viewing their sustainability activities as a given that speaks for itself.

To preserve their competitive edge, these companies have to be watchful and make themselves heard in the presence of companies exploiting any sustainability activities as campaign topics.

However it is not all about the "if" of communication, but also about the "how", for example: what signals does an advertisement send? What ideal of beauty does it promote? Who does it possibly latently discriminate against?

And finally, the question is also: which stakeholders the company actually communicate with and in what form? Is it able to handle criticism by NGOs or the media, for example?

3.3.2.5 The Relationship with Employees

The way a company treats its employees is a key aspect of its sustainability performance.

Equal opportunities for all employees, and their promotion irrespective of gender, race, age, disability, or sexual orientation, as well as employee development according to their abilities, the highest possible level of security, and fair pay are key points.

The term "human resources" is highly problematic in this context, since it does not appreciate employees as "humans" but instrumentalizes them in their function as "resources."

3.3.2.6 Corporate Philanthropy

In addition (not as a substitution) to all previous topics, philanthropy is a good opportunity to give back to society and strengthen solidarity, whilst also communicating company values.

On offer – depending on the sector of business, company, and brand values – is the whole range of cultural, social, and educational engagement.

Particularly if philanthropic activities are realized locally, they strengthen regional ties and people's identification with the company.

3.3.3 Social Diversity and Differentiation

All of the above social aspects and topics significantly contribute to sustainable development. They also offer an opportunity for businesses to develop a distinct profile, which allows them to differentiate themselves from competitors, to strengthen the brand and motivate employees and external stakeholders.

This can only work, however, if all sustainability activities are aligned with the company's values and brand(s). As there is biodiversity, so there should be a diversity in approaches to social sustainability among companies.

Weleda, presented in the following pages as a case study, took the step of developing a distinct sustainability strategy aligned to its identity: Weleda's corporate sustainability strategy was developed using a highly participative approach, to gain the widest possible input and generate buy-in from internal and external stakeholders [2]. To ensure an international perspective and sensitivity towards regional differences, six countries were chosen to participate in the strategy's development: The three base markets of Switzerland, France, and Germany and additionally the Netherlands, the UK, and the USA. The strategy was intended to represent stakeholders, and stakeholders were to feel represented by it: only in this way would the strategy come to life.

3.4 CASE STUDY – WELEDA: A VALUE-ORIENTED BUSINESS

The Weleda Group, based in Arlesheim near Basel, Switzerland, is a leading manufacturer of complementary medicines, anthroposophic pharmaceuticals, and natural and organic cosmetics. It employs more than 2000 people around the world and sells its products in over 50 countries across five continents.

Because of its unique roots, social values form a central component of Weleda's corporate culture. The company was jointly created by the founder of anthroposophy, Rudolf Steiner, and a Dutch-Swiss doctor, Ita Wegman, in 1921. Rudolf Steiner, a twentieth-century reformer, advocated a holistic view of humanity that reveals its influence in several areas of modern life such as the Waldorf schools, organic cosmetics, a heightened environmental consciousness, and biodynamic agriculture. Weleda, in turn, was founded as a business based on the principles of ecological, social, and economic responsibility, and its pharmaceuticals and holistic cosmetics were intended to benefit mankind. There is, therefore, a direct link between what is today known as *corporate sustainability* and Weleda's core business.

There are seven core Weleda values:

1. Fair treatment of customers, partners and suppliers.
2. Management–employee relations in a spirit of partnership.
3. Extensive quality.
4. Combining natural and spiritual sciences in research and development.
5. High environmental standards.
6. Ethical and value-creating business practices.
7. Cultural diversity as an inspiring force.

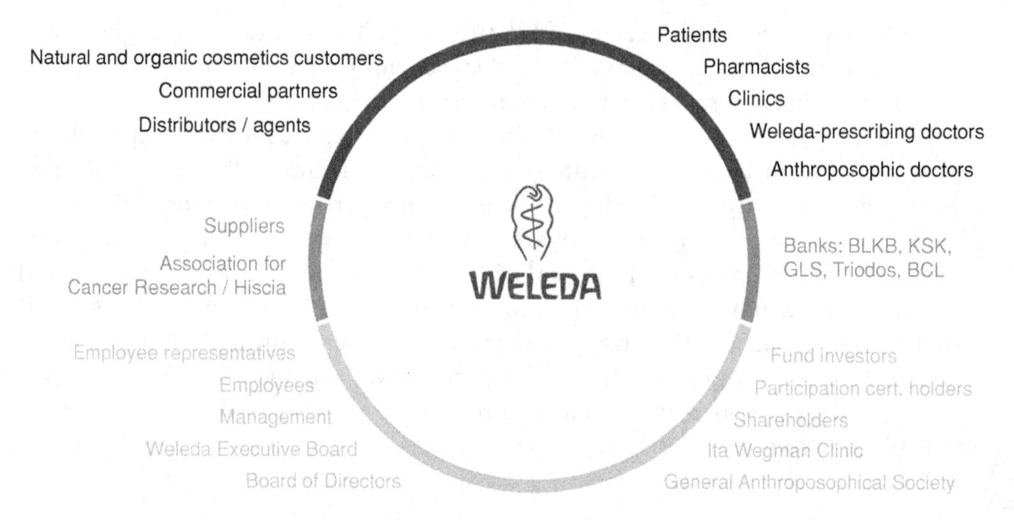

Figure 3.1 Weleda: A productive social organism – key stakeholders.
Source: Reproduced with permission from © Weleda.

Employees, suppliers, customers, and patients of Weleda are all considered part of a productive social system comprising a number of different stakeholders (key stakeholders are listed in Figure 3.1). Initiatives that strengthen the weaker stakeholders are expected to benefit the system as a whole.

Such initiatives have typically arisen independently across the various locations in which Weleda is present, reflecting the diverse social reality of the Weleda Group. They were first coordinated across the Group in 2010 with the formulation of Weleda's corporate sustainability strategy: as wide as possible a variety of inputs from different countries and departments were considered in the process in order to reflect this diversity, with more than 700 people participating from both inside and outside the company. Under the motto of "creating a meaningful footprint," Weleda then defined seven fields of action, encompassing not just social, but also environmental and economic responsibilities, with goals to be achieved within the next five years (see Figure 3.2).

Three of these fields of action hold particular relevance for this chapter:

The first two: *environmentally friendly raw materials and biodiversity* and *fair trade* relate directly to its core business and can be combined under the term *ethical sourcing*, a key principle of Weleda's approach to its raw materials. Third, *space for working and living* is a field in which Weleda's corporate values are especially apparent in its particular approach to its own employees – and a field in which, somewhat unusually, customers are known to have high expectations of the company. We will present some examples of initiatives ongoing in these three fields. We will then take a brief look at the area of corporate philanthropy: this field was deliberately not included in the corporate sustainability strategy since it

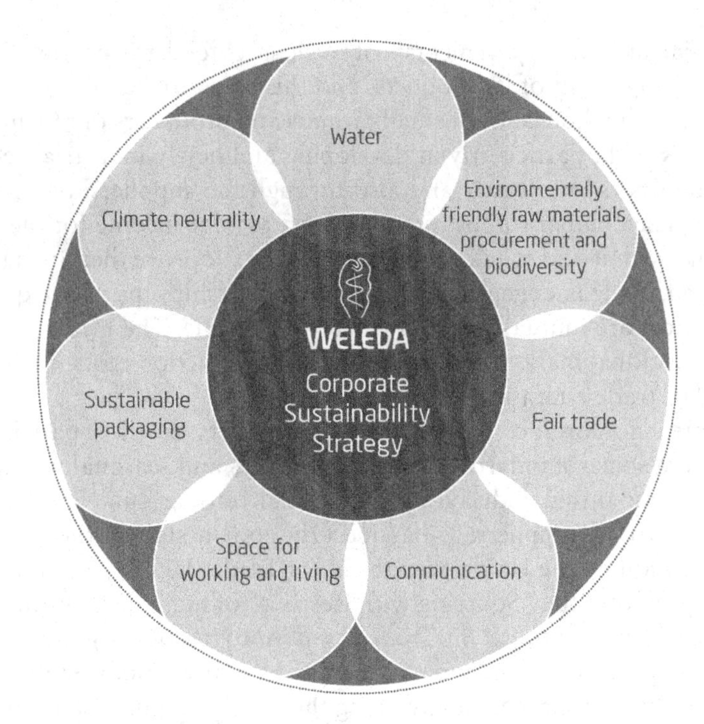

Figure 3.2 The Weleda corporate sustainability strategy.
Source: Reproduced with permission from © Weleda.

does not relate directly to Weleda's core business, but it nevertheless offers a useful insight into the Weleda approach.

We will conclude by widening the scope of this chapter: Weleda's approach to social responsibility is closely tied to the principle of sustainable economic development as an essential long-term guarantor of its initiatives. The final section will examine this link, rooted in the company's holistic worldview, and look at some of the initiatives within this field.

3.4.1 Ethical Sourcing

The concept of ethical sourcing combines two fields of action within Weleda's corporate sustainability strategy (see Figure 3.2): *fair trade* and *environmentally friendly raw materials procurement* (or, in other words, sourcing of certified organic and biodynamic raw materials) *and biodiversity*. Weleda currently operates over 50 partnerships around the world for the cultivation and wild collection of raw materials. In doing so, great attention is paid to holistic value creation for its partners, which not only includes financial security, but also social progression.

The "social quality requirements" of Weleda extend to the entire production and value-adding chain of its products, and this serves to reinforce the company's commitment to the thousands of small farmers and producers of raw materials that it collaborates with, particularly in developing and newly industrialized countries. Weleda supports them directly and also through the suppliers and distributors it works with. The company aims to guarantee a regular income for the people who work for and with it and to take active measures to improve their living conditions.

To this end, Weleda engages in long-term partnerships; purchase quantities and minimum prices are mutually agreed wherever possible. The prices are determined in part by the global market, but chiefly by real production costs. Only in this way is genuine fair trade guaranteed.

It is of great importance to Weleda that suppliers do their part in complying with minimum social standards for their employees and seasonal workers. For this reason, the company introduced its own social trading guideline back in 2006, which applies to all suppliers. It has recently gone a step further: in October of 2011, Weleda joined the Union for Ethical BioTrade (UEBT), a non-profit organization that promotes the "Sourcing with Respect" of ingredients from biodiversity. Weleda will apply the *Ethical BioTrade Standard* to its sourcing practices, promoting biodiversity conservation and the sustainable use of biodiversity, and making sure that benefits are shared equally along the supply chain. UEBT has a very high expertise in the field of biodiversity and fair trade and its standard provides a model that Weleda considers to be fully in line with its own values.

3.4.1.1 Wild Sourcing of Arnica Montana*: Combining Social and Environmental Benefits*

Weleda sources its medicinal plants mainly from biodynamic cultivation in its own gardens and from partner companies around the world. However, a small proportion of plants are sourced from certified wild collections. Amongst these is arnica (*Arnica montana*), sometimes jokingly referred to as the "patron saint of muscles and bruises." It plays an important role as an ingredient in medicinal ointments, and has also been used for over 80 years in the manufacture of arnica massage oils, one of Weleda's most important products.

A rich source of arnica can be found in the Apuseni Mountains in Romania, where there are also some of the highest human settlements in the Carpathians. The interests of the local communities are uniquely interdependent with the existence of the arnica flower in their region. *Arnica montana*'s native habitat depends on sustained agricultural use: it grows on meadows used by the local population to make hay (see Figure 3.3). If the meadows were abandoned, the flower would be displaced by shrubs and trees within 2–3 years. Conversely, the traditional, semi-nomadic lifestyle of the communities – whose cultural traditions include the crafting

Figure 3.3 *Arnica montana* flowers grow on meadows used for making hay. Apuseni Mountains, Romania.
Source: Reproduced with permission from © Weleda.

and playing of a regional version of the alpine horn – would be under threat without an adequate income, forcing them to seek work in the major cities.

There is a tradition of harvesting medicinal plants in Romania, a country which, during the 1970s and 1980s was one of the leading producers of medicinal and aromatic plants in the global market. Cultivation and wild gathering are organized and monitored by state-run enterprises. Following the end of the communist system, the market collapsed and it has only recently begun to gradually recover. But recovery has also brought with it some negative effects: noticeable over-exploitation of the medicinal plant population in the wild has created a list of more than 20 endangered species of plant. Arnica can, for example, react very sensitively when grazing land is suddenly used intensively. This can happen as a result of fertilization, which can strongly alter the nutrient balance in the soil. Arnica also reacts sensitively to intensive harvesting. As a result, the arnica sources disappear.

Since 2005, Weleda has been working alongside the World Wide Fund for Nature (WWF) to counteract the depletion of *Arnica montana* in the Romanian Carpathians. In partnership with a cooperative, the local population and a Romanian agriculturalist, the company has implemented a collection method in the region of Cluj-Napoca that takes the interests of all parties into account, in addition to protecting natural resources.

Protection through Use

Five villages in the Apuseni Mountains: Gârda de Sus, Arieşeni, Scărişoara, Albac and Horea participate in the Weleda project. Around 450 people own the meadows on which the arnica grows, and approximately 500 persons participate in collecting the flowers.

Drawing on its past experience with a similar project in the Vosges Mountains in France, Weleda, together with its partners, has created a reliable method of collection and conservation that ensures sustainability for the arnica medicinal plant. A specified number of flowers are collected by hand and the grazing areas are used exclusively for extensive livestock farming, and remain unfertilized. Annual training is required to implement this strategy. The farmers that harvest the arnica – who predominantly live off livestock and forestry – are trained by members of the local cooperative. The farmers receive a stable additional income as a result of the annual collection every June. The accompanying scientific "monitoring" of the plant population (detailed observation of the arnica population on precisely surveyed land allotments) provides insights leading to further improvements and delivers valuable data on long-term collection and harvest management.

Local Drying: Better Quality for Weleda – More Independence for Its Partners

The commission of a newly built drying plant in 2007, pre-financed by Weleda, is of great significance to the arnica project. It was built entirely from the abundant local timber, a sustainable natural resource which is also used for heating. Weleda has been able to contribute by providing a continuous exchange of knowledge and experience in the drying of medicinal plants to the cooperative.

The drying plant, which provides Weleda and its partners with high quality arnica flowers, has made the farmers and the members of the cooperative more independent, helping them take a further link in the value creation chain into their own hands.

3.4.1.2 Rose Cultivation in Turkey: Sustainable Sourcing with Social Benefits

The refined scent of the damask rose provides the Weleda Wild Rose range of natural and organic cosmetics with its unique fragrance. According to legend, rose oil was worth its weight in gold in ancient times: the essence of the rose flower was considered so valuable by the rulers of Persia and the Arab World that they were prepared to pay any price for it. The reason for this lies partly in the work that it takes to produce it: three million flowers must be hand-picked to produce a single liter of essential rose oil.

In collaboration with Sebat, a local distiller, Weleda has created a pioneering project on the high plateau of the Isparta province in Turkey, where 300 farmers have switched their operations to sustainable organic rose cultivation since 2001. This is now the world's largest organic rose project: one third of the worldwide crop for the sought-after fragrance extract, rose absolue, is organically produced here for Weleda.

Fair trade and organic quality are important pillars of the Weleda sourcing strategy, and they are integral to the Weleda brand. The cooperation between Weleda and the Sebat rose distillery provides a good example for a sustainable business practice: Weleda guarantees purchase of rose absolue and, as a result, receives a consistently high level of quality in plentiful supply. Thus over the last ten years, a stable partnership with many advantages has been created between Weleda and local farmers: since the switch to organic cultivation, they are guaranteed a secure and regular basic income and their harvests are partly pre-financed. In addition, the switch has provided benefits to the environment as well as to their own health, as no pesticides or insecticides are used.

The Weleda rose project in the Isparta region began in 2001 when the local rose oil distiller Sebat, a Weleda-supported Turkish consultant for organic cultivation, and 30 small-scale farmers wanted to convert their scented rose fields into organic farms (see Figure 3.4). During the first few years, there was a great deal of skepticism from the local population and a number of conventional farms. There was barely any knowledge of organic farming methods without the use of insecticides, pesticides, and synthetic fertilizers, and so the on-site assistance from the consultant was able to make a considerable difference. As the organic cultivation methods proved successful, the initial doubts started to fade. Within a few years, what began as an initiative to promote sustainable rural development became a partnership with an effect on the entire region. Since then, over 300 small-scale farmers have switched to organic rose cultivation: the farming area is now around 200 hectares and the annual harvest now brings in around 600 tonnes of organic flowers. From that total, approximately 700 kilograms of valuable rose absolue is

Figure 3.4 A flower picker works through the bushes of an organic rose field in Isparta, Turkey. *Source:* Reproduced with permission from © Weleda.

Figure 3.5 *Strophantus kombé* flower growing in the wild, Malawi.
Source: Reproduced with permission from © Weleda.

collected for Weleda. Thanks to further expansion of organic cultivation methods, the fields and plants in the area are healthy and the rose farmers receive fair prices for their produce and a guaranteed market for their high-value products. Weleda can depend both on them and a secure supply of a premium natural product.

Weleda and the Sebat distillery have used the income from the project to finance a village kindergarten for 50 children from the farms, between the ages of three and six years. Moreover, there is financial support for student projects, and scholarships are awarded to those who wish to continue research into the organic cultivation of roses. This academic support contributes to the future sustainability of the project.

3.4.1.3 *Access and Benefit Sharing:* Strophantus Kombé *Wild Collection in Malawi*

Strophantus kombé, also known as dogbane (Figure 3.4), has been employed as an effective remedy in the treatment of heart conditions. The plant has long been known to hunters in Malawi as a useful arrow poison. Its medical benefits are said to have been discovered by accident in 1859, when a member of David Livingstone's expedition to the region was cured of pains in the chest after some of the poison came into contact with his tooth brush. Based on a principle of intellectual ownership by the indigenous population, a Weleda project in Malawi for the wild collection of strophantus is producing earnings that go into social programs.

Strophanthin, a pharmaceutical contained within the plant, has been employed successfully in Germany since 1905 for various heart problems. According to anthroposophic studies, strophanthin can also protect against *angina pectoris* and reduce psychovegetative cardiovascular disorders. The principle is that the seed fat acts on the activity of the heart, but also emits a warming effect, thereby providing a holistic healing effect for the patient. Hence, strophanthin is an important medicinal product for Weleda. The plant also plays a small but significant role for a section of the population of Malawi, where over 90% of the population work in agriculture. The harvest time of the *Strophanthus kombé* plants lies between the harvest high seasons of tobacco, tea, and sugar cane, and thus offers an additional income for most of the country's harvesters.

In 2007, Weleda could no longer provide strophantin-based products as a result of problems in raw materials procurement: the amounts harvested from collection in the wild were well below the annual demand of 1500 kilograms. Harvesting the long, unwieldy fruit of *Strophanthus kombé* requires particular strength, skill, and experience, since they grow on vines located several meters above the ground. The plant's behavour and its potential for use as a crop had not been fully researched and, in addition, more and more forest in Malawi was being destroyed by the devastating use of fire to clear forest in order to create space for agricultural planting – taking with it the wildlife required by the *Strophanthus kombé* plant for pollination.

Beginning in 2008, Weleda and a partner company, TreeCrops Ltd, developed a concept that provides a new basis for traditional wild collection, supplementing it with cultivation as a crop. Since 2010, the project has been co-financed by the German Society for International Cooperation (GIZ) in the form of a public–private partnership (PPP). In coordination with local authorities and communities, various forest areas in southern Malawi and the Lower Shire area were assigned and mapped for exclusive conservation and cultivation of *Strophanthus kombé*. The idea is to aim for a sustainable collection method that qualifies for organic certification. In addition, some of the cleared forest aisles are being used as cultivation areas for a possible semi-domestication of strophanthus. Initial tests are promising.

An access and benefit sharing mechanism (ABS) has been a core component of the project from the start: based on the principle that the local population should share in the profits gained from knowledge of the medicinal qualities of strophanthin, which was discovered thanks to its traditional use as an arrow-tip poison. Five per cent of the sale price of *Strophanthus kombé* from TreeCorps to Weleda is returned to the village communities as royalties based on intellectual property rights. The communities decide what the money should be used for: in the village of Lumvira, for example, it was spent on repairs to the village water fountain.

Since the start of the project, TreeCrops Ltd. has worked with a regular team of collectors who have been specifically trained for this task and who receive a harvester identification card upon completion of their training. The training includes the identification and proper collection of the right species and the correct length of time for drying and collecting the seeds. Most training coaches can be found in the expansive network of villages. They receive 2 dollars per day for their work. The minimum wage in Malawi is 1 dollar per day. Only fully trained collectors are allowed to work in the Weleda depots.

As of 2011, several hundred collectors and suppliers have been involved in the project, with an increasing number of committees being formed in the villages to collaborate with TreeCrops and to take advantage of the value of the forest as a source of income.

3.4.2 Employee Policy

It is a stated goal that anyone who works at Weleda should find it a place of professional and personal development. As with the concept of *salutogenesis* that underlies the Weleda approach to health, the company's approach to its employees emphasizes personal responsibility, a core principle of anthroposophy.

Services for employees are diverse and vary between sites. Within the confines of this chapter, we will present an overview of initiatives underway in the four largest Weleda locations in terms of employees: Germany, Switzerland, France, and the Benelux region, which together encompass 75% of the Weleda workforce (see Table 3.1). We will then present a closer look at the Weleda Generation Network, an acclaimed recent social initiative in Germany.

3.4.2.1 *Raising Awareness and Personal Development*

Four of the seven Weleda values described in Section 3.4 are directly connected to social responsibility: fair treatment of customers, partners and suppliers; management-employee relations in a spirit of partnership; ethical and value-creating business practices; and cultural diversity as an inspiring force. To ensure that employees are familiar with them and with the philosophy from which they are derived, Weleda offers management seminars and training programs. One example

Table 3.1 Weleda employees in the four largest locations as of 31 December 2010.

	Weleda Germany	Weleda Switzerland	Weleda France	Weleda Benelux (Netherlands and Belgium)
Total number of employees[a]	966	230	385	119
Total number of male employees	288	102	117	43
Total number of female employees	678	128	268	76
Nature of employment				
Full-time employees	680	156	297	38
Part-time employees	286	74	88	81
Of which are trainees, apprentices, marginal employees, volunteers	68	12	37	0
Employees by age group				
Up to 30 years of age	223	54	76	11
31 to 50 years of age	566	122	254	71
51 to 60 years of age	152	43	49	27
over 60 years of age	25	11	6	10

[a]Total number for Weleda Group: 2259.
Source: Reproduced with permission from © Weleda.

is the "Weleda Identity and Basic Values curriculum," which is open to managers and prospective managers as well as new employees, and which provides a basic introduction to Weleda's anthroposophic corporate culture.

In addition, various career programs, management courses, and anthroposophic seminars are offered to encourage professional and personal development, supplemented by internal coaching for managers and employees. These courses place an emphasis on personal responsibility and social competence.

Employees are encouraged to think about ecological, social, and humanitarian issues. To this end, courses and presentations for raising awareness are regularly offered at all locations, often in cooperation with external partner organisations. Awareness has been known to translate into concrete initiatives: for example, in the case of an employee mobility plan in the Netherlands that was awarded with an innovation prize by the Dutch Employers' Association in 2010 for encouraging the reduction of car travel on the part of employees. The plan both reduces harmful emissions and encourages a healthy lifestyle.

3.4.2.2 *Promoting Health*

Weleda takes care to provide comprehensive measures for the health and safety of employees at work. Work canteens with organic and wholefood dishes contribute to employees' wellbeing, and workplaces are analysed and assessed to ensure that they are safe and ergonomic. In 2010, under the project name Salus (Latin for "wellbeing"), Weleda Switzerland launched a work safety programme in cooperation with Suva, the Swiss accident prevention organisation. The first employee training sessions, as well as up-to-date, documented safety guidelines, have resulted in a reduction in the number of work-related accidents, and also in technical and organisational optimisations.

3.4.2.3 Integration and Diversity

Weleda brings together people with different requirements, cultures, and ages. Mutual respect, support, and a willingness to continually learn from one another are standards that are seen as integral to the corporate culture, and cultural diversity is embraced as an inspiring force. Weleda France and Weleda Germany (where more than half of Weleda's employees are located) have signed public charters in their respective countries that promote diversity and integration and help prevent discrimination in the company.[1] At Weleda Netherlands, around 7% of employees in 2010 were people with a disability. In France, there is an integration project for deaf employees (*A deux mains* – "with both hands").

3.4.2.4 Career and Family

Weleda's holistic philosophy finds expression in an open, family-friendly approach towards its employees: flexible working hours and childcare facilities aim to facilitate the combination of career and family life. There is support for mothers, part-time opportunities for older employees, and paid leave to provide home care for relatives. In cooperation with a private social counselling service, Weleda Switzerland offers support to employees in difficult situations. Weleda Germany, the company's largest subsidiary, has been operating its own daycare center run on Waldorf principles since 1998, and since 2005, employees who become new parents have been planting birch saplings in the Weleda medicinal plant gardens for each new child. In France, in August 2010, the children of Weleda employees started to attend the company's own new crèche, Coeur d'enfants, located directly next to the Huningue site.

3.4.2.5 Weleda Generation Network

A recent Weleda initiative in Germany, the Generation Network, is notable for its unique approach to balancing career and family and to dealing with demographic change, for which it received the 2010 innovation prize of the

[1]www.charte-diversite.com and www.charta-der-vielfalt.de.

Bundesverband Deutscher Unternehmensberater (Federal Association of German Management Consultants).

The idea for the Weleda Generation Network emerged in 2004 in the context of an audit process conducted by the German *Hertie Foundation*, and it was then further developed as a project. Conceived as a bridge to connect generations, the project was designed to improve intergenerational dialogue and counteract loneliness on the part of older people, and to satisfy practical, everyday needs at the same time.

The Generation Network is sustained by a volunteer team comprised of active and former employees, and its services are made available to all Weleda employees, retirees, and their family members (explicitly including employees without children). The services on offer include help in and around the home and in the garden, petsitting for vacations, help in emergencies of all kinds, purchases, computer support, driving services, technical assistance, preparation for retirement, emergency childcare, and childcare for company parties and conferences. Compensation for these services is agreed on an individual basis with the people involved. It is possible to exchange services, pay for them, or to receive them as voluntary services.

The network also helps find partners for leisure activities and organizes groups with shared interests for purchasing goods. In addition, it provides an umbrella for retirees to organize lectures on relevant topics such as, for example, patient's rights regarding life-prolonging measures in the event of coma. A shopping service is organized by the Weleda Generation Network and offered by external service providers. Internships for young people with disabilities also grew out of this project, and it is now possible for employees to buy high quality fresh fruit and vegetables, breads, and cakes on the company premises in Schwäbisch Gmünd. Additionally, the Weleda Generation Network views itself as a point of contact for soon-to-be retirees. To make the start of retirement easier, which can represent a profound change in a person's life, the network organizes joint activities and discussions.

A contact person within the company coordinates these activities, and offers and requests are published in a public display case and by e-mail.

Weleda views the network as a modern form of solidarity: as flexible as possible and as binding as necessary. By supporting employees and making it easier for them to balance their careers and private lives, it serves to make an effective contribution to health and wellbeing at the workplace. In the meantime, companies from around Germany have shown interest in the benchmark for the Generation Network model. Weleda regularly invites those who are interested to share ideas.

3.4.3 Corporate Philanthropy

Unlike its sourcing and employee policies, philanthropic activity is not as much connected to Weleda's core business as it is to its values. Nevertheless, it is viewed as an important extension of social responsibility that can complement other activities. Following the principle of diversity that forms a core Weleda value, philanthropy is encouraged to take a variety of forms in the individual countries in which Weleda is present, typically as a response to needs discovered through interaction with local communities.

Recent examples include a cooperation between Weleda's Italian subsidiary and the Milanese hospital Mangiagalli, on a project to promote safe childbirth in the Esmeraldas region of Northern Ecuador; the doubling of a cash prize awarded to Weleda in Germany in recognition for the Generation Network project, and its donation to *wellcome*, a Hamburg organization for the support of young families; and the contribution by Weleda in Belgium to self-supporting biodynamic cultivation of medicinal plants at a village community and shelter for Sri Lankan refugees, founded by a Belgian activist in Sevapur, India.

In individual cases, there can be a more direct connection to Weleda's business: as part of its raw material procurement, Weleda both promotes and, in some cases, demands the creation of social projects in various countries which create secure jobs, regular income, cooperation, and trust. Weleda is convinced that this is the only way to make an effective, credible contribution to local stability and global development. In the former Soviet Republic of Moldova, for example, a farming cooperative supplies Weleda with organic lavender. A partnership based on fair trade and the support of organic cultivation with Weleda expertise has gained an additional dimension after it became apparent that access to medical care for the local farming families was poor, which led to an endowment for the construction of a medical and welfare center in the town of Mencini.

3.4.4 Economic Sustainability and Value Creation

However socially responsible a company may be, its long-term benefit to all stakeholder groups and to society as a whole is inherently tied to continuous stability and solid growth. For this reason, the corporate aims of the Weleda Group are centered on the sustainable creation of value: Weleda views economic success as a prerequisite for fulfilling its social responsibility to customers, employees, internal and external investors, suppliers, and partners, as well as society and the environment.

Sustainable business requires responsible trading within clear management structures and binding rules of conduct; this also includes the sensible handling of risks. In this respect, in 2010 the Weleda Group developed comprehensive concepts and tools that are being established across the company, including a Group-wide code of conduct and a control and risk management system.

3.4.4.1 Value-Added Statement

Conducting business with a focus on social responsibility means returning profits to the value-added cycle and thus providing impetus for further internal and external

Table 3.2 Weleda Group value-added statement, 2010.

Value-added statement 2010	2008 in millions of CHF		2009 in millions of CHF		2009/2008 change in %	2010 in millions of CHF		2010/2009 change in %
Employees	143.2	99.6%	157.2	93.0%	+9.8	148.1	97.1%	-5.8
Public authorities	1.7	1.2%	2.1	1.2%	+23.5	3.3	2.2%	+57.1
Operating result	-5.4	-3.8%	4.3	2.5%	-179.6	-4.4	-2.9%	-202.3
Shareholders	0.0	0.0%	0.7	0.4%	–	0.0	0.0%	–
Lenders	4.3	3.0%	4.8	2.9%	+11.6	5.5	3.6%	+14.6
Total added value	143.8	100.0%	169.1	100.0%	+17.6	152.5	100.0%	-9.8

All information given at the prevailing rates at the time specified

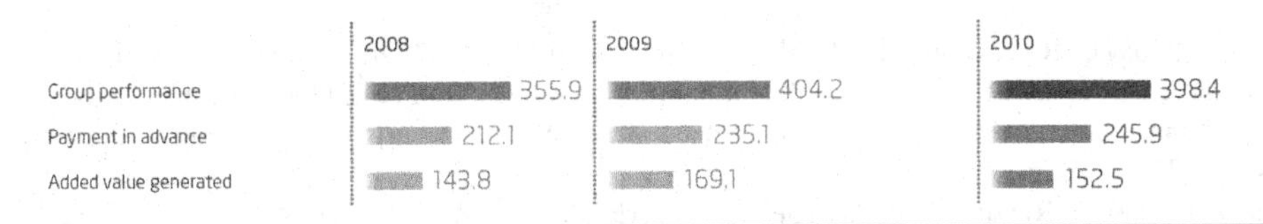

Source: Reproduced with permission from © Weleda.

development. This cycle must be quantifiable, which is why the Weleda Group employs a *value-added statement* as an important tool: it shows how Weleda's business actions create value for the company and its stakeholders. Unlike the profit and loss statement, which is based on the viewpoint of the owners, the value-added statement explains the contribution of the Weleda Group to private and public income. It shows the expense at which Weleda achieved its company performance and how the generated added value was distributed. The value-added statement for 2010 provides an example (see Table 3.2).

In the 2010 financial year, company performance increased, after exchange rate adjustments, due to a significant growth in turnover of 13.5% to CHF 398.4 million. After deducting advance payments, the added value of the Weleda Group was CHF 152.5 million. The added value per employee of the Weleda Group was CHF 79 000 an increase, after exchange rate adjustment, by around 1% compared to the previous year. A seven-figure amount was spent on donations and similar contributions to associated social institutions in 2010. This amount is included in the advance payments. The greatest proportion of the added value, 97% (CHF 148.1 million), was used for paying Weleda employees, which are viewed as co-entrepreneurs involved in the business process. The public authorities received 2.2% (CHF 3.3 million) in the form of taxes. 3.6% (CHF 5.5 million) of the added value was allocated to interest to lenders. Around 1.8% of the non-distributed added value was available for financing the future growth of the company.

3.5 CONCLUSIONS

This chapter has shown that the social footprint of an organization is complex as it involves assessing the impact on suppliers, customers, employees, as well as all other stakeholders. It is thus to no surprise that much work has been done on measuring environmental impacts, but none on social impacts. By using our company (Weleda) as a case study, we have shown how cosmetic companies can create social value for suppliers, partners, and employees. Key to creating such social value is economic growth (profits), without which none would be possible.

RECOMMENDED READING

&EQUITY (2009) CSR – thenewAdded Value. Reputation und Nachhaltigkeitsimage von Marken und Unternehmen im Kontext des LifestyleofHealthandSustainability (LoHaS), Hamburg, Germany.
Organic Monitor (2010) Strategic Insights: CSR & Sustainability Initiatives in the Beauty Industry, London, UK.
Weleda Sustainability Report 2010, Arlesheim, Switzerland.

REFERENCES

[1] Organic Monitor (2010) Strategic Insights: CSR & Sustainability Initiatives in the Beauty Industry, London, UK, p. 7ff.
[2] Weleda Sustainability Report 2010, Arlesheim, Switzerlandp. 4.

4

Ethical Sourcing of Raw Materials

PART 1: ETHICAL SOURCING – THE GIVAUDAN APPROACH

Remi Pulverail

4.1.1 THE BUSINESS CASE FOR ETHICAL SOURCING

For Givaudan, the world's largest purchaser of raw materials in the fragrance industry, the business case for ethical sourcing could not be clearer: securing more than 100 strategic supply chains, now and into the future, without compromising the environment, the livelihoods of producers, or profitability, is business critical.

Ethical sourcing is about taking vital steps to improve supply-chain transparency and ensure full traceability of naturals used in perfumery, to satisfy both customer requirements and increasing consumer demand.

Most of the natural ingredients Givaudan uses are right at the heart of the fragrance business. It has a responsibility to make sure that it secures the volumes required, while safeguarding the jobs and revenues of tens of thousands farmers around the world who make a living selling these crops. Without rose, jasmin, lavender or patchouli, perfumery would lose a significant part of its distinct heritage and beauty. Although not an easy task, the company is committed to getting even more involved, at the source, to ensure an ongoing supply of these materials.

The company sources specialist natural ingredients, such as vanilla, tonka beans and benzoin, the production of which are concentrated in less developed countries such as Madagascar, Venezuela and Laos. The challenges are clear, and are closely linked to the global evolution of agriculture. Increased pressure on food commodities, and decreasing soil productivity – a consequence of an over usage

Sustainability: How the Cosmetics Industry is Greening up, First Edition. Edited by Amarjit Sahota.

of pesticides – are contributing factors. Another key threat is the loss of a young generation of potential farmers from agriculture, which is why the company works hard at keeping motivation high for growing strategic niche crops.

With raw materials as one of the five 'pillars' of its corporate sustainability programme, Givaudan works towards an ultimate aspiration of 'sourcing raw materials in ways that preserve the environment, stimulate the development and wellbeing of communities, and safeguard the efficient use of precious resources'. It is a vision that comes from the leadership of the company and is carried into the daily operations of the purchasing department.

Givaudan has taken the industry lead to work on strengthening fragile supply chains, through encouraging agricultural best practice and ensuring that the communities involved benefit long term. The company's Givaudan Innovative Naturals programme, in particular, is focused on providing its perfumers and, in-turn, its customers, with a long-term, sustainable natural ingredients supply.

By developing supplier partnerships, seeking alliances and external expertise, the company not only makes sustainable natural ingredients available to its perfumers – it is also discovering exciting new-to-perfumery natural products which have never before been incorporated into fragrances.

There are currently five different Givaudan Innovative Naturals ethical sourcing programmes in Australia, Venezuela, Laos, Madagascar and the Comores Islands. Each is built on two basic principles:

1. sustainable development, which involves preserving the environment; for instance, by harvesting at the appropriate times and replanting for the future; and,
2. fair trade, which focuses on improving quality of life for local people and reducing the pressure to leave the countryside, to find work elsewhere, by ensuring a sufficient income.

Each programme is a long-term commitment for Givaudan and has a dedicated budget. However, the ultimate goal for each initiative is that they become self-sufficient so the programme can be expanded and new sustainability-related initiatives financed.

4.1.2 MAKING ETHICAL SOURCING A REALITY

Ethical sourcing can rarely be tackled by one company in isolation. Givaudan has worked closely with all its raw materials suppliers to ensure that they understand the standards it needs to fulfil its philosophy. In February 2010, the company became a signatory to the United Nations Global Compact, which aims to align policies and business practices with universally-agreed values concerning human rights, labour, environmental protection and anti-corruption. It is a commitment that reverberates throughout the company's supply chain. A Global Vendor Expectations document notifies all raw materials suppliers that they too need to comply with the United Nations Global Compact's ten principles.

The company is committed to operating to the sustainable business model that it believes is vital to the future of the industry. It has its own 'Key Principles of Sustainability' document, approved by the Executive Committee in December 2011, and publically available on www.givaudan.com to enable anyone to read about the formalised sustainability framework it operates under.

All Givaudan's manufacturing sites participate in the Supplier Ethical Data Exchange (Sedex). Sedex operates an extensive and secure, web-based database of information based on four pillars: Labour Standards; Health and Safety; Environment; and Business Integrity. Givaudan also has a target of having 200 of its own top 400 suppliers join the platform by 2015, and has appointed an external company to assist it in performing the required Sedex audits.

4.1.3 WORKING WITH CUSTOMERS

Whilst the consistent and secure supply of high-quality, and sometimes new-to- perfumery, materials is the most significant benefit to Givaudan and its customers, the Innovative Naturals programmes have also provided a useful foundation for discussions about ethical sourcing and sustainability in general. Indeed, several customers have welcomed exploring an alternative, more sustainable, way of formulating fragrances. For example, by increasing the quality and reducing the quantity of raw materials used in fragrance formulations, perfumers are able to deliver memorable fragrances whose purity consumers relate to and desire. The company's range of fully natural and Ecocert certified fragrances, (a French environmental body devoted to sustainable development and the certification of organic agricultural products) and biodegradable formulations, also provide customers with the opportunity to consider marketing such products.

The Innovative Naturals programmes are, first and foremost, about securing raw materials supply. Major multinational agricultural companies focus their supply-chain investment on high-volume commodities such as palm, soya, cocoa and coffee, leaving Givaudan, a purchaser of 'niche' crops, with no option other than to safeguard its own interests. Customer marketing opportunities and claims which arise out of the use of these ethically-sourced raw materials are important, but are always secondary to this.

Ethical sourcing programmes aim to build as prosperous a supply chain as possible so customers are not able to have exclusive involvement in any of the programmes. They can, however, capitalise on Givaudan's expertise in ethical sourcing to design a specific long-term natural supply-chain initiative of their own if they wish to do so.

4.1.4 BUILDING SUPPLIER PARTNERSHIPS

The success of the Givaudan Innovative Naturals programmes is based on strong and long-term supplier partnerships. These provide local growers with revenue security, as well as technical and productivity assistance, in exchange for a commitment to the supply of perfumery materials, and the protection of the unique places in which they grow.

Identifying the right supplier partner is a complex, and often long, journey for the procurement team. The relationship must work for both sides; for the community involved, and for the business, to be sustainable long-term – which is always the objective. In many cases, it is also necessary to liaise with local government officials, a process that, in itself, can be arduous and requires diplomacy and lengthy negotiation.

Increased transparency between the producers and Givaudan has been the foundation of the successful partnerships. Only by gaining a deeper understanding of each stakeholder's constraints can scarce resources be managed effectively. Details of these, and other supplier partnerships the company has developed, are given in the case studies below.

4.1.5 SECURING THE FUTURE OF BENZOIN IN LAOS

Benzoin resin is an ingredient which is widely used in prestige fine fragrances. It is a gum that is extracted from *Styrax tonkinensis* trees by villagers in the remote Phongsaly and Houaphanh provinces in northern Laos. Farmers harvest the resin by climbing up the tall *Styraz tonkinensis* tree, cutting the bark and collecting the resulting benzoin resin 'tears' which seep out – a very labour-intensive process which is dependant on the skills and availability of the local population (Figure 4.1.1).

People living in these villages understand the value of education. Although primary education facilities in this area are generally adequate, the number of secondary schools is insufficient and so many parents are forced to move their families or send their children away to school, threatening social cohesion and the future workforce for benzoin collection.

Faced with this threat of a rural exodus from the benzoin sourcing area, Givaudan entered into a unique partnership with a Laos-based agricultural products and supplies company. This company is the interface between the gum collectors from the local communities, the importer and processer of the gum, and Givaudan. The partnership formalises a commitment from both sides to ensure a sustainable

Figure 4.1.1 Benzoin resin oozes from the bark of the *Styrax tonkinensis* tree in the Phongsaly and Houaphanh provinces of Laos. The dried 'tears' are collected by villagers and provide additional income for these rural communities.
Source: Images courtesy of Givaudan SA.

supply and trade in benzoin. It involves various education initiatives to develop the communities at the centre of benzoin gum extraction. For example, Givaudan financed the building of two secondary schools to provide secondary education for more than 100 teenagers from villages across the region.

Givaudan and its business partner are also participating in a programme regarding the long-term management of the forest ecosystem to increase *Styrax tonkinensis* tree planting and improve benzoin collection methods. The partnership has developed new ways of extracting and treating the gum, leading to increased volumes of the raw material being used, securing more income for local communities (Figure 4.1.2).

The partnership has studied the possibility of complementing the cultivation of the *Styrax tonkinensis* forests with facilities for the supply of other fragrance raw materials. Since 2008, three new products have joined the Givaudan perfumers' palette: Beeswax Absolute Organic (2008), Ginger Red Oil Organic (2009) and Cinnamon Bark Oil Orpur® (2010) – each providing additional revenue for the local benzoin collection communities and innovation to Givaudan perfumers as these materials have never been used before.

Figure 4.1.2 Approximately 40 000 people, mostly farmers, live in about 150 villages in the Phongsaly and Houaphanh provinces of Laos where Givaudan and its local partner, a Laos-based company, arrange the benzoin gum collection from local communities.
Source: Images courtesy of Givaudan SA.

The beeswax story is particularly interesting. Villagers source honey for their own consumption from wild bee nests perched high on forest trees. They climb the trees to collect the nests and then separate the honey from the crude beeswax.

The beeswax, which was traditionally used to make candles for Buddhist ceremonies, is typically washed, filtered and moulded into big shapes before being sold in rough pieces in Vientiane, the capital of Laos. Now, Beeswax Absolute Organic is extracted instead, with 200 kg of Beeswax resulting in 4 kg of Beeswax Absolute. The final product, used by Givaudan's perfumers, is certified organic by Ecocert.

4.1.6 TRACING THE ORIGINS OF ETHICAL VANILLA IN MADAGASCAR

Madagascar is the source of over 80% of the world's vanilla crop. However, there are no large plantations; the vanilla supply chain is uniquely based on a rural structure of villages in the north-east of the country.

The growing, harvesting and curing process, which involves many steps, demands a high level of expertise from the local growers to maintain the highest possible quality. Each vanilla flower – the plant is part of the orchid family – is pollinated by hand. After the harvest and during the curing process, all the beans are carefully hand-selected to eliminate those of lower quality.

The number of intermediaries, and lack of transparency in the supply chain between local growers and buyers like Givaudan, poses risks to both parties: inferior

Figure 4.1.3 Vanilla is a member of the orchid family that needs the human touch to produce its fruit; every flower is pollinated by hand, the green pods harvested and nurtured under the sun for up to six months to produce the rich dark brown bean.
Source: Images courtesy of Givaudan SA.

quality of production for buyers; and economic uncertainty and lack of visibility for local growers (Figure 4.1.3).

To mitigate such risks, Givaudan is working directly with Malagasy vanilla farmers to help improve the yield and quality of their vanilla crop. Working in an exclusive partnership with a local supplier that has been curing and exporting Malagasy vanilla since 1911, it now has a dedicated supply from a number of local villages, offering complete control over the growing, harvesting and curing.

Not only does sourcing vanilla in this way help ensure future supplies, but the use of technology and organic farming practices have also secured traceable and certifiable supplies of vanilla – both topics of considerable importance to customers.

Each field has a Global Positioning System (GPS) location to guarantee traceability, and all vanilla sourced from this ethical vanilla sourcing programme is certified organic by Ecocert. Participating farmers have the security of knowing that they have a guaranteed market for their product over at least five years.

As well as implementing a new patented curing process for the vanilla, an agronomist has worked with local farmers to optimise vanilla growing and harvesting practices to ensure maximum curing is achieved. Such initiatives include preventing a specific root fungus with the potential to destroy vines, and the proper maintenance of the 'host trees' on which the vanilla vines climb. The company also works alongside the Malagasy villagers to reduce their reliance on rice imports by intensifying their own rice production – a main staple of their diet.

In addition to paying quality premiums, and offering vanilla planting and curing advice, the programme also includes renovating or building schools to ensure the future education in local communities. Givaudan is working, on an exclusive basis, with 14 villages representing a couple of thousand farmers. It has provided such support to eight villages so far, and plans to support all 14 by 2014.

4.1.7 MOHÉLI PARTNERSHIP REDISCOVERS YLANG YLANG

Ylang Ylang, quite literally means 'flower of flowers'. Its heady scent is described as 'sweet, slightly spicy floral', and its oil is highly prized by perfumers. It is a tropical flower that is mainly harvested in the Indian Ocean islands of Madagascar and the Comores (Figure 4.1.4).

The essential oil is produced locally through steam distillation of the flowers. This produces different grades, with high density fractions being used in fine fragrance and lower density fractions used in consumer products and soaps.

Figure 4.1.4 Ylang Ylang is a fragrant treasure with a heady, slightly spicy scent that evokes the vibrant colours of summer. The project in Moheli, will safeguard traditional production techniques and secure premium quality Ylang Ylang essential oils.
Source: Images courtesy of Givaudan SA.

Over the years, the quality of Ylang fractions has deteriorated. It has become common for the best grades, Ylang Extra, to be adulterated with lower grades in order to cope with the price differences and unbalanced demand between the different fractions.

To prevent this rich material from disappearing completely, it was necessary for Givaudan to restore transparency throughout the supply chain and establish a more direct link with local producers. Additionally, it was of utmost importance to follow up and control each step, from the flower picking to the distillation of the oil, and ensure that quality would again be at the heart of the full process.

To ensure best practice in cultivation and secure a guaranteed quality of Ylang Ylang, the company has developed an exclusive partnership in Mohéli, the smallest island of the Comores archipelago. Working with a local producer with many generations of experience in producing and exporting aromatic plants from the Comores, it is investing in specific distillations of Ylang Ylang. The project has many dimensions that include investments to support the expansion of children's education through the purchase of school supplies; the installation of new distillation stills to improve the quality of the oil, and to increase the wood and water efficiency of the production process. In addition, each farmer has been made aware of the urgency of safeguarding wood resources – nurseries have been established in each village participating in the programme to offset the usage of wood for the distillation, and to ensure that planting takes place to replace the trees that have been harvested.

Cultivation practices are critical to the quality of the Ylang Ylang essential oil. Givaudan has also been helping collectors to understand the importance of only picking the yellow flowers that are in full bloom, leaving the other flowers to mature and yield good quality essence. A picker can harvest up to ten kilos of flowers a day which, critically, must then be distilled within two hours of their harvest or they will lose their fragrance (Figure 4.1.5).

> "All our work at Mohéli was about rediscovering the real fragrance of Ylang Ylang, which we had lost" explains Rodrigo Flores-Roux, senior perfumer at Givaudan. "The quality had gradually degraded and the floral splendour was nowhere to be found. Until I finally smelled the real flower, I was like a painter who was blind to a colour. Discovering this flower's scent has brought new tonalities to my olfactory palette."

4.1.8 EQUIPMENT LOANS SUPPORT SUSTAINABLE SANDALWOOD PRODUCTION

Sandalwood is fundamental to the customs and culture of the Indian subcontinent, and is used from cradle to cremation in the Hindu and Buddhist faiths. As one of

Figure 4.1.5 Cultivation practices are critical to the quality of the Ylang Ylang essential oil. A picker can harvest up to ten kilos of flowers a day which, critically, must then be distilled within two hours of their harvest or they will lose their fragrance.
Source: Images courtesy of Givaudan SA.

the most precious raw materials in perfumery, it has been renowned for hundreds of years for its subtle and refined woody scent.

Since 2001 Givaudan has preferred to use Australian sandalwood due to the deforestation and quality issues that threaten the Indian supply. In June 2007, it signed a partnership agreement with a producer of pure Australian sandalwood oil.

The partnership created a sustainable supply of the 'Indigenous-certified' grade of sandalwood oil, which is harvested by people from the Aboriginal communities in Golden Outback, the southern part of Western Australia. Historically, only one fifth of native Western Australian sandalwood was collected by Aboriginal people, due to their lack of equipment, vehicles and harvesting licences.

As a result of the partnership, a special fund has been set up to finance equipment, which helps indigenous harvesters to continually improve harvesting techniques.

Equipment purchases such as loaders, grabbers and forklifts have enabled more efficient 'whole tree' utilisation, while other harvesters have built site offices, refurbished their vehicles or purchased new generators.

In addition, the company is paying a premium on each kg of essential oil sourced from this channel, a process which is audited by an Australian independent certification body.

4.1.9 PROTECTING BIODIVERSITY AND TONKA BEAN SUPPLY IN VENEZUELA

Tonka bean is a precious, wild-grown natural raw material used in various luxury fine fragrances due to its surprisingly profound warm and sweet smell, reminiscent of caramel, almond and tobacco all rolled together. Venezuela is one of the two main areas where tonka beans are collected globally.

Conservation International's Conservation Stewards Program (CSP) and Givaudan have been working together since 2007 to conserve the Caura Basin of the Amazonian forest in Venezuela, to support local livelihoods and a secure supply of tonka beans.

Under a new agreement, signed in 2011, the initial protection area of 88 000 hectares will be expanded by a further 60 000 hectares over the next three years – securing, in particular, Sarrapia trees from which tonka beans are collected, as well as important habitats for local wildlife such as the American Tapir, Jaguar, Spectacled Caiman and Yellow-spotted Amazon River Turtles, among others.

The CSP agreement promotes community-based action to protect biodiversity and strengthen the link between local tonka collectors and international buyers. Through the agreement, the quality of life of the 64 families of the Aripao community is enhanced. Local communities receive technical and financial assistance from Givaudan and CSP, through a local NGO, in exchange for their commitment to preserving the forest and its precious flora and fauna.

This is leading to a better harvesting, drying and storage process of the tonka beans, thus improving the quality of supply. The company is also funding the maintenance of harvesting routes, which increases the volume of beans collected in high production years. With the new agreement signed, it is also now exploring the development of a new supply chain for a second non-timber forest product – copaiba balsam.

The higher output of beans that results from the agreement enables the local community to increase its revenue. The initiative is helping the local community to improve the organisation of the tonka and copaiba supply chain. A small percentage of the money received for the beans is held in a bank account, which is managed by the community. Funds from this can be used as a line of interest-free credit to help people invest in equipment, seeds or in promotion of small businesses.

Finally, as part of the tonka project, Givaudan has also developed an exclusive product, Tonka Roasted Bean, which is certified organic by Ecocert. Tonka Roasted

Bean brings new intonations of cocoa powder to the perfumers' palette for use in fragrance creation. It also creates extra revenues for local collectors.

4.1.10 IS NATURAL SUSTAINABLE?

Current market demand for 'natural' is high, which leads to the question – can the world sustain our demand for it? Natural ingredients have inspired perfumers for centuries – the earliest recorded uses of scented products for religious or cosmetic use were all extracted from nature itself. However, the most sustainable approach to perfumery does not always mean sourcing from nature. The fragrance industry needs to use both natural and synthetic materials to continue to supply the modern, continuously expanding, global market. Nature can only supply a limited amount of essential oil, sustain a certain number of trees, and there is only enough land for so many fields of flowers – so how do you balance the need to grow to supply the fragrance industry with the need to feed people?

Not everything that can be smelled in a flower or fruit can be extracted, sustainably or otherwise. Some scents are changed forever by the process. Various technologies have been developed to capture the essence of nature by other means and without the need to involve the plant itself, merely identification of the scent molecule within it. Some of the most delicate and familiar scents would not be available in fragrance without technology – the scents of lilac for instance, or lily-of-the-valley.

Science provides a way to bring novelty and innovation into perfumery. For example, developing new extraction and purification methods leads to more efficient use of limited supplies; re-using waste products can reduce the use of new ones; or the invention of cleaner processes can help protect the environment.

Biosciences are providing new opportunity to bring innovation in naturals to perfumery. Working within the guidelines from the United Nations Conference on Trade and Development, Givaudan is seeking opportunities to bring never-seen and never-smelled natural and innovative ingredients to the perfumer's palette – where there is always an appetite for new or novel ideas.

4.1.11 CONCLUSION

Ethical sourcing is a very complex process that takes into account all the arguments and puts forward a compromise to find the best solution. There are very many variables including that of climate, the limits of supply, and the work that is necessary to inspire best practice for the environment in, often, very different cultures. The truth is that ethical sourcing is a commitment that businesses must enter cautiously and responsibly. The agreements made will dramatically affect the livelihoods and environments of those involved. With the drive for more and more transparency in the beauty industry, it is inevitable that ethical sourcing will become more common place. But to be truly effective it must be genuine – which means it must meet clear business objectives, be founded on sound basic principles that benefit all parties involved, and be a long-term commitment that reflects a philosophy and not a marketing claim.

PART 2: INNOVATION AND ETHICAL SOURCING – BERACA'S EXPERIENCE

Filipe Tomazelli Sabará

4.2.1 INTRODUCTION

The word innovation derives from the Latin *innovare*, an expression that is known to date back to 1500 and means 'to renew or change'. In other words, an innovation can be considered anything that produces novelty as a consequence. Ideally, innovation should be a current that flows through the organisational chart of every institution, regardless of hierarchical levels. In order to deliver an improved outcome, all members of the chain must be engaged in making improvements focused on reaching innovative results. It should also change the way decisions are made, creating positive impact, making room for new choices and values.

However, innovation has become one of the greatest paradoxes of our time. Living in a world where everything needs to be faster, more efficient, more productive and preferably cheaper, technological improvements have made it possible for the industry to reach outstanding results in time-efficient, cost-effective ways.

On the other hand, in the midst of such a crucial effort towards development, we come across the fact that our planet is no longer capable of handling the supply for our level of production, and terrible environmental consequences are already a devastating reality. Moreover, social inequality is still a great weakness in our society and, regarding ethical sourcing specifically, there is still much to be done.

Therefore, some crucial questions emerge. How to innovate in a sustainable and ethical manner? Would we be willing to accept the challenge of innovating in a way to invest in ideas that aim to optimise the entire value chain? Will we be open to develop truly long-lasting relationships among all stakeholders, including companies, consumers and communities under nature conservation and social enhancement principles? Are we open to true transparency?

It becomes clear that organisations should be committed to identifying the planet's vulnerabilities, helping foster socio-environmental innovation through sustainable development in a global context, not just looking at short-term interests.

After 1996, when the United Nations Conference on Trade and Development (UNCTAD) launched what is now known as the BioTrade Initiative, the innovative concept of production was created – with the transformation, and commercialisation, of goods and services derived from native biodiversity under the criteria of environmental, social and economic sustainability. However, ethical sourcing means the assumption of new responsibilities by organisations for their labour or contracting of services regarding much more than just simple human rights practices within its supply chain; they also have to be conducted through the building of transparent long-term relationships based on trust and integrity through new environment and social enhancement values that promote real benefit-sharing, while also taking into account product quality standards. Thus, will sustainability be a true form of innovation in the sense that it represents a new way of living, doing business and perpetuating the world and life, as we know it?

Is there any light at the end of the tunnel?

Fortunately, the cosmetic industry has already been for many years at the forefront of the sustainability discussion. There is increasing demand for green, natural, organic and sustainable products, with a large number of companies investing in such products. It is realised that the more the industry embraces the concept and launches new products, the more it helps increase the number of consumers who are willing to buy them. Also, new technologies are emerging, quality is improving, and perpetuation is made possible.

However, as has already been said, adopting simple environmentally-friendly practices is not enough anymore. Sustainable development requires the complete involvement of the whole production chain, including the sustainable sourcing of nature-derived ingredients: an extremely complex process from extraction to distribution.

It has become necessary for the industry to supply products made from raw materials that are from transparent and traceable sources in strict compliance with global socio-environmental standards. The interdependent factors – environment preservation, economic vitality and healthy enhancement of communities – become inevitable.

4.2.2 CHALLENGES RELATED TO ETHICAL SOURCING

Taking this context into account, one of the first practical steps towards establishing a sustainable production chain is guaranteeing the responsible collection of natural resources coming from all kinds of biomes and biodiversities which give birth to our industry's materials supply, including virgin forests and the people with traditional knowledge and the interdependence related to it.

However, the biggest challenge appears to be to transform traditional to scientific knowledge, and then on an industrial scale, whilst following sustainable processes.

Transparent relationships must be established prior to harvesting, and contracts must be based on benefit sharing: taking into account the correct access and use of genetic resources and intellectual property rights. Companies must also invest in local development, which includes improvements in infrastructure, training, handling and certification of raw materials; companies therefore have to communicate and work closely with other stakeholders, such as universities, non-profit organisations, the media and the local government.

Researchers at the University of Maryland in the United States estimate that the income generated by forest resources should correspond to approximately US $1.1 trillion per year [1]. At the same time, according to a study by Dr. Philip M. Fearnside at the National Institute of Amazonian Research (INPA), it would be necessary to spend US$ 3 trillion per year to control the negative effects of global warming as a consequence of deforestation [2]. The conservation of natural resources must also be combined to improve the quality of life of the local population.

The best way to reach this balance is via neo-extractivism, which makes use of specific technology and respects local people's identity and lifestyle, applying fair prices and compensation policies for environmental services such as clean water and soil, carbon sequestration and reforestation. In this sense, the challenge is clearly related to ethical sourcing involving the identification of sustainable activities that can be performed by local communities on an industrial scale within the environmental capacity. Furthermore, each extractive product has individual ecological, economic, social and ethnic-botanical characteristics that require special attention in terms of labour absorption, manufacturing expertise, stock management, market development, shelf-live evaluation and socio-economic organisation. Preservation of natural resources is only possible through traceability and monitoring of the entire production chain, from extraction to delivery at the customer end.

Other challenges involved in ethical sourcing are promoting responsible and sustainable practices within the finished consumer product, developing structures to meet sustainability-related demand, implementing policies to reduce social and environmental impacts and reaching global markets interested in promoting sustainable products.

4.2.3 BERACA AND THE BIODIVERSITY ENHANCEMENT PROGRAMME

Beraca was founded in 1956 in São Paulo, Brazil, by the late grandfather (Ubirajara Sabara) of this author. Since 1978, the company has been led by his son and the father of this author, Ulisses Sabara. The author of this contribution is proud to represent the third generation of Beraca Sabara in a year (2012) that the company celebrates 56 years of existence. Beraca is now a leading provider of natural and organic ingredients from the Amazon Rainforest and other Brazilian biomes to more than 40 countries. The company has seven factories in Brazil and offices in Sao Paulo, New York and Paris. Our experience has been considered a success story of an organisation that is working today, but also putting in practice the new and challenging concepts that are anticipated for the future. Beraca has been investing in the development of sustainable ways of doing business through its belief that socio-environmental responsibility is essential to ensure the success of future generations.

Environment responsibility is a strong value that has driven all the company's activities, developing a corporate culture based on principles such as equity, transparency and ethics. Through partnerships, certifications and awards, Beraca reaffirms its concrete initiatives towards sustainability. It seeks to educate its clients on the importance of acquiring legitimate natural active ingredients from certified sources, whilst making commitments to global biodiversity principles. Beraca also assesses social and economic impacts involved in the development of its products and services, placing emphasis on building ethical and transparent relationships with partner communities, investing in skilled labour and state-of-the-art technology. The active ingredients sourced from the Amazon Rainforest are refined at Beraca's branch in Ananindeua, located in the metropolitan area of Belém, state of Pará. Research and development of products is conducted in partnership with regional universities, contributing to the value of the traditional knowledge of the Amazonian population and the equitable sharing of benefits. Meeting the economic, environmental and social concerns of different groups and communities is one of Beraca's main goals.

In order to make sure the work is done properly, Beraca created its Biodiversity Enhancement Programme in 2000. The programme aims to ensure sustainability and traceability of supply of raw materials from Brazilian biodiversity. Beraca operates as a bridge between scientific organizations and local communities in the introduction and dissemination of technologies and products, providing fair benefit-sharing along the supply chain. The programme also contributes to regional development, selective waste collection and residue reduction, to preserve the world's largest rainforest: the Amazon. The company currently benefits around 1600 people in 101 community centres by assisting in their organisation process, increasing environmental awareness and promoting knowledge transfer. Besides adding value to products, these initiatives also increase their market competitiveness.

With Beraca's support and guidance, needy communities have found alternative sources of income to ensure a better quality of life for their families and preserve the Brazilian biodiversity. All Beraca projects aim to promote the sustainable use of natural resources, while respecting individual rights and traditional knowledge. The chain of custody begins from research and product development, enhancing the knowledge of riverside communities whilst adding new knowledge and technologies.

With a wide network both in the national and international market, Beraca helps to create demand for the communities through partnerships with small to large cosmetic companies interested in a fair and sustainable supply chain. All local projects are developed through international concepts of sustainable development and communication.

4.2.4 WORKING IN PARTNERSHIP WITH LOCAL COMMUNITIES

4.2.4.1 Organizational Support and Business Management

Beraca discovered that in many local communities, earnings were restricted by a lack of organisation and business inexperience. To overcome this, the company promoted the creation of business micro-cooperatives. It began to hold regular meetings with leaders of associations and communities to discuss details of the production process. The meeting discussion points are recorded in the form of minutes, signed by the community leaders and company representatives.

The topics addressed in the meetings include community organisation (the role of each community member), leadership, the importance of taking minutes, agreements and invoices, and the consistent administration of payments for the supply of products. All member communities have a current account to receive their earnings and minimise theft, thereby promoting good administration of income for the benefit of the community.

Beraca also encourages local community inhabitants to recognise the importance of their work as representatives of the Amazon region, and teaches how the government, Sebrae (Brazilian Service of Support to Micro- and Small-Sized Enterprises) and NGOs can contribute to the development of the community's activities.

4.2.4.2 Beracom

In 2010, Beraca held the first Beracom: Beraca and Community Leaders Dialogue in the Amazon region (Figure 4.2.1). The event brought together community leaders from five Brazilian states, many of whom had never left their communities before. Representatives of Sebrae, Ideflor (State Institute for Forestry Development in the state of Pará) and non-government organisations were also present.

The purpose of the event was to give guidance to supplier communities on aspects that drive commercial demand, administrative issues related to the purchase and production of raw materials, and the actions of the Biodiversity Enhancement Program. The meeting also addressed financial matters (billing and payment forms,

Figure 4.2.1 Coomac leader José Martins, during his speech at the first Beracom.
Source: ©Beraca.

as well as deadlines), concepts of organic farming and sustainable harvesting, and access to the Brazilian genetic heritage. A visit to Beraca's factory and laboratory located in the city of Ananindeua, state of Pará, was also arranged.

Beracom 2011 was also attended by members of the Ministry of Agriculture, the Department of Agriculture of the State of Pará, the State Department of Finance of Pará, the regional consortium Bolsa Amazônia and NGOs Instituto Vitória Régia, Fundação Cáritas and GIZ (German International Cooperation Agency) (Figure 4.2.2).

Beracom was intended to become an annual event. It aims to promote interaction and exchange of experience between community leaders, by demonstrating Beraca's transparency in its relationships with local communities, and seeks to improve its partnerships through the analysis, comments, criticisms and feedback of all the delegates.

4.2.4.3 Guaranteed Demand

Beraca signs supply agreements that guarantee the purchase of products and respective deadlines. In addition, the company can make advance payments, via a microcredit system, ensuring working capital for the communities. Requests for advance payment enable Beraca to track information on each supplier through an identification code, grant the advance payment to those with approved registration and

Figure 4.2.2 2012 Beracom participants.
Source: ©Beraca.

no debts, and render accounts within a maximum of 30 days after authorising the request for advance payment.

4.2.4.4 Training and Qualification

Beraca works on the co-creation of new harvest techniques and good agricultural practices to ensure the best performance of the community and contribution to environmental preservation. Topics addressed include the importance of hygiene and the use of appropriate recipients for the storage of products, organisation and production logistics (the right harvesting time of each product), quality control (appropriate storage of the fruit so that it does not spoil quickly, avoiding loss and waste) and organisation of the harvesting (harvesting fruit and seeds only, not stumps or stones).

Beraca instructs communities on the organisation of product storage prior to sale (to prevent theft), the use of specific tools to extract raw materials without damaging the species (*copaiba*, for instance), and accident prevention (not exceeding the weight limit of boats, precautions to avoid snake bites and the importance of wearing protective boots in dense forests).

Beraca also trains communities in matters relating to labour legislation, international codes and standards such as the SA 8000, which guarantee the best conditions for all workers: facilities, toilets, working methods and principles of health and safety. Overall, concepts related to human and environmental development are taught, such as the importance of education for children, valuing customs and history (keeping traditions alive), waste management (recycling) and the risks of contamination through incorrect waste disposal.

Organic Certification

With organic certification of products, such as Ecocert, Beraca adds value to the community. Certification also ensures all production processes are reviewed, confirming good environmental and social practices.

To raise awareness of the importance of certification in the community, Beraca holds meetings to explain what organic products are, their distinguishing features in relation to non-organic products, and their importance in the current market. Beraca also teaches communities about the need to sort and store organic products appropriately, using special recipients (provided by Beraca) and special techniques that do not use agrochemicals or alter the products. The certification agencies perform annual audits to check the facilities and ensure the communities are meeting all the requirements of organic standards.

4.2.4.5 Crop Diversification

Partnering with Beraca allows communities that once were tied to just one crop to work with crop rotation throughout the year, diminishing the effects of seasonality and increasing income.

4.2.4.6 Promotion of Partnerships

Beraca stimulates partnerships between local communities and the company's national and international clients (such as L'Occitane, Estée Lauder, Unilever and L'Oréal), as well as the government and NGOs.

4.2.4.7 Selection of Suppliers

Beraca does not work with local suppliers that do not operate according to the company's requirements. All suppliers must be committed to the environmental and social principles established by Beraca. The company, in turn, helps its suppliers in the development of evaluation and screening procedures to ensure all the procedures are carried out correctly.

4.2.4.8 Traceability

The entire production process, from harvest to product delivery to clients, is traceable to ensure the preservation of natural resources and biodiversity. Beraca ensures the control and follow-up of the origin and any changes to the product, throughout the production chain.

4.2.4.9 Placement on the Market

Through partnerships with Beraca,communities that were previously isolated gain exposure on the national and international market.

4.2.5 SUCCESS STORIES

4.2.5.1 Cooperative Coopemaflima (Island of Marajó, Pará)

Before the partnership with Beraca, the Coopemaflima community depended mainly on fishing and subsistence farming. During the wet season, fishing was less lucrative, causing a marked decline in the community's earnings.

In view of this, Beraca began to promote the collection of the andiroba seeds that are carried downriver and deposited on the beaches of Marajó. Past generations used these seeds to create an artisanal wax that was used as a repellent. The large number of andiroba trees in the region had led the local authorities to consider the seeds as waste; they were burning down hundreds of trees every day to 'clear' the beach for tourists (Figure 4.2.3).

Beraca realised that collecting andiroba seeds was an opportunity to create an alternative source of income. The company helped the community to establish as a cooperative in order to sell products, get organic certification, as well as learn techniques for collecting the seeds, extracting the oil and selling it to the cosmetics industry. Beraca currently purchases around 500 tons of seeds per year, benefiting around 1000 community members.

The next steps were to obtain organic certification, to start production and to provide training to the community in oil manufacturing. In partnership with Fondation L'Occitane, the headquarters of Coopemaflima was built, with funds coming from the state government for the processing plants. The land was purchased by Beraca and the construction work was funded by Fondation L'Occitane.

The community members also began to produce candles from andiroba oil and Marajoara ceramics for L'Occitane Brasil. The partnership with Coopemaflima was responsible for Beraca receiving the BBC World Challenge Award 2007 and the SEED Awards 2009; the latter was created by the UN for entrepreneurship

Figure 4.2.3 Women from Coopemaflima collecting Andiroba seeds at the seashore.
Source: © Beraca.

in the area of sustainable development. Beraca, Coopemaflima and other partner companies are currently working on new projects to continue building sustainable businesses.

4.2.5.2 Community of Panacauera (Igarapé-Miri, Pará)

Before the partnership with Beraca, the community's main income was from fishing and the cultivation of açaí. Both activities can only be carried out from August to December. As a result of the monoculture of açaí, other local species were cut down to increase the land area for planting açaí. Moreover, because the community earned its income from a single crop, the limited income during the rest of the year led some members of the community to cut down trees to sell wood to ensure their own survival.

To resolve the problem, Beraca introduced buriti as an alternative crop, which is harvested from January to July. The community of Panacauera was already established as a cooperative. Beraca provided training sessions on forest preservation, harvest techniques, organic certification and the manufacture of by-products of buriti, such as confectionary and oils. Beraca also provided the community with equipment for drying the fruit and producing buriti oil, flour and confectionary; this helped generate a new source of income for the community.

In partnership with POEMA, a German NGO that is affiliated with the environmental department of University of Pará, the results obtained over three years of project are as follows: preservation of açaí and buriti trees and other local species; diversification of production; and increased income, with a consequent improvement in the community's quality of life.

Figure 4.2.4 Beraca's and Kiehl's team with Nazarezinho local community.
Source: © Beraca.

4.2.5.3 Community of Nazarezinho (Igarapé-Miri, Pará)

Responsible for the cultivation of açaí, the community of Nazarezinho was trained by Beraca on business management, organic certification, good handling practices and post-harvesting practices. Beraca also promoted a closer relationship between the community and Kiehl's – a division of L'Óreal – which uses Beraca's açaí extract in its products.

Kiehl's representatives have visited the community of Nazarezinho on a regular basis (Figure 4.2.4). In 2010, the company held a competition named 'Vision of a sustainable forest' for children and young people of the community. The participants were asked to draw a picture that reflected what a healthy, well-protected tropical forest meant to them. The three best drawings were rewarded with a cash prize, whilst the winning drawing was featured on the label of the product 'Açaí Damage-Protecting Toning Mist', manufactured by Kiehl's.

At the In-Cosmetics 2011 trade show for cosmetic ingredients, Beraca and Kiehl's promoted the projects of sustainable Açaí crops, which are cultivated by communities like Nazarezinho (Figure 4.2.5).

4.2.5.4 Ilha das Onças (Ilha das Onças, Pará)

The community works with andiroba, murumuru and pracaxi crops. Beraca trained the community in good production practices and organic certification, and provided

Figure 4.2.5 InCosmetics showcase – from plant to product. Beraca and Kiehl's sustainability case.
Source: © Kiehl's.

a greenhouse for drying of murumuru seeds and a shed for storing the fruit before its transport. To make the organisation of documents and the receiving of e-mails easier, Beraca donated a laptop to the community.

4.2.5.5 Farmer's Association of the Community of Santa Luzia (Tomé-Açu, Pará)

Before teaming up with Beraca, the community was supplying pepper and cupuaçu crops to the food industry. Beraca presented the possibility of also supplying cupuaçu oil to the cosmetics industry, which would increase the community earnings and the number of people employed.

Beraca introduced the Community of Santa Luzia to the Inter-American Development Bank (IDB) and the Ministry of Agriculture, enabling the community to be included in the Pará Rural programme. This provided resources for the acquisition of an agro-industry plant for the community. Beraca also donated an industrial stove and a depulper for the production of traditional confectionary made of cupuaçu, encouraging sales at regional trade markets.

In 2008, in partnership with ABIHPEC (Brazilian Association of the Cosmetic, Toiletry and Perfumery Industry), Beraca took eight international journalists to visit the community of Santa Luzia. The visit resulted in articles published in international media, including a feature on the cover of GCI Magazine: a publication for American executives in the cosmetics industry.

4.2.5.6 Community of Caetés (Bragança, Pará)

The Community of Caetés used to practise crop rotation with buriti and andiroba, but their activities were related only to the sale of fruits. Beraca donated a greenhouse for the natural drying of fruits, and taught the members new techniques for peeling the seeds (through a heating process and the use of a special tool). Beraca's R&D team also trained the community in the production of buriti and andiroba oil for cosmetic applications, using the equipment available at the cooperative headquarters.

4.2.5.7 Community of Jubim (Salvaterra, Pará)

Clay is the main product for this community; since it is a mineral, it cannot receive organic certification. Beracatook the Ecocert certification team to visit the community, which was granted a fair trade certification that attests the traceability and transparency of the entire production chain. Beraca also invested in infrastructure to ensure the correct and sustainable extraction of clay, donated computers and a printer to the Department of Environment and the local school, provided a gate for the main office of the local association and promoted sharing of the benefits with land owners and neighboring communities.

4.2.6 WHAT IS YET TO BE ACHIEVED

The future is uncertain and many challenges lie ahead, as we enter into a new era with completely new rules and ways of living while doing business. While we as companies are producing in state-of-the-art plants, with high yields and increasing volumes, reaching levels of technology never imagined before, there are still people living under the poverty line, struggling not only to live, but also to survive. Take the healthcare market for example: through technology we are capable of saving people from diseases, extending life by months, years or even decades, however there are still over a billion people dying of hunger or without access to a single meal a week. Does it make any sense?

New opportunities are appearing. For instance, banks like UBS from Switzerland are investing in what is known as 'impact investment'; this is starting to play a new and important role, whereby they help new entrepreneurs in socially weak territories to develop new kinds of business, whilst investors look for new ways of doing business that provide a better return on their investments but also help develop poor areas, especially in developing countries. Operators at the base of the pyramid are becoming the best partners for making business profitable, as the social enhancement factor is starting to gain more and more strength. Therefore we have to be prepared for a crucial change in every aspect of life as we know and live it.

We are still in the middle: a transition moment between the way we used to see development, innovation and business, and the new and challenging reality where companies and individuals are now being obliged to think not only on their present needs, but also on how to innovate in a way to fulfil current demand while thinking about the perpetuation of the planet and its people.

Quality and cost is still a challenge when developing green and safe products. The consumer is still not clear about where companies want to go and what they will offer as a solution to their demanding and sophisticated tastes.

As players in the beauty and cosmetic market take part in developing, launching and making products for the urgent demand – but also, on the other hand, become knowledgeable about the reality – we have to be lucid and brave enough to 'spread the word' to mostly naive consumers. In this sense, information becomes essential and communication has to play its role, shouting about the urgency for change.

Globalisation and social media must use their strength to involve and educate people, making it possible to unite and involve consumers and the whole chain of custody so they sit at the same side of the table and agree on the new reality that we need to live in and start to create new dimensions.

In this sense, certification also has to start playing a new and more harmonised role so consumers are able to truly understand and differentiate what is what; so they are no longer fooled by the hundreds, if not thousands of logos.

Governments must finally come to agreement regarding access and benefit-sharing practices and create new and global rules that are effective, but at the same time do not affect development.

In general, a more effective and efficient participation of the final consumer in the whole process is definitely needed for a truly transparent way of doing business, promoting real equitable sharing of benefits as well as a new model for the actual fair trade proposal, with the participation of all members of the process.

Being directly active in the cosmetics industry, either as entrepreneurs, executives, investors or consumers, we have to understand the reality of the big picture and be part of the change.

4.2.7 CONCLUSION

The model of innovation through partnerships between companies, communities and consumers, under nature conservation and social enhancement, is a modern one.

It promotes a true vision of the planet and identifies potential vulnerabilities in the growing demand; it helps create projections and then produces true innovation resulting in changes that can promote social and environmental sustainable development globally.

> Sustainability means, long lasting. We have to be able start and start over, to adapt and readapt, to think and rethink. To know and to deeply understand who we are and what we want to become, for years to come and others to follow.
>
> —Filipe Sabara.

REFERENCES

[1] COLIPA. COLIPA Activity Report. 2008. Available from: https://www.cosmeticseurope.eu/downloads/136.html.

[2] Union for Ethical BioTrade. About the Union [Internet] 2011. [Accessed 2011 December 17; cited 2011 December 17] Available from: http://www.ethicalbiotrade.org/about/index.html.

5

Biodiversity in the Cosmetics Industry

Eduardo Escobedo and Rik Kutsch Lojenga

5.1 INTRODUCTION

Biological diversity (biodiversity) – the variety of life on Earth – forms the web of life on the planet at three different levels: ecosystems, species and genes. Biodiversity underpins the functioning of the ecosystems and the services they provide to sustain life on Earth and human well-being, such as food and fresh water, clothing, fuel, health and recreation, air quality and climate regulation, waste treatment, and protection from natural disasters to name a few. Table 5.1 illustrates how the ecosystem services enjoyed by people depend on both the diversity (quality) as well as the amount (quantity) of genes, species and ecosystems.

All businesses depend on, and impact, biodiversity. The cosmetics industry in particular depends on biodiversity as it provides a source of innovation and raw materials. Typically, these raw materials are used after undergoing some type of transformation, producing oils, surfactants, emulsifiers and other ingredients. Table 5.2 gives some examples of cosmetics ingredients derived from natural resources that are dependent on biodiversity.

Sustainability: How the Cosmetics Industry is Greening up, First Edition. Edited by Amarjit Sahota.

Table 5.1 Relationship between biodiversity, ecosystems and ecosystem services.

Biodiversity	'Quality'	'Quantity'	Services (examples)
Ecosystems	Variety	Extent	• Water regulation • Nutrient cycling • Recreation
Species	Diversity	Population	• Natural ingredients • Design inspiration • Fuel, timber, fibres
Genes	Variability	Number	• Scientific discovery • Disease resistance • Adaptive capacity

Source: Adapted from TEEB in Business and Enterprise (2012).

In its annual report 2008, Cosmetics Europe – the Personal Care Association identified respecting of scarcity of natural resources, reducing biodiversity damage and developing resource efficient product life cycles as among the primary challenges for the cosmetics industry [1].

The recognition of the importance of biodiversity by some companies has led to the creation of programmes to prevent biodiversity loss by enhanced business practices that create positive economic and social incentives for the sustainable use and conservation of biodiversity.

In 2007, the Union for Ethical BioTrade (UEBT) was launched with the commitment of its members to gradually ensure that their sourcing practices promote the conservation of biodiversity [2]. In 2008, a group of companies from the flavour, fragrance, cosmetics, personal hygiene and food processing industries launched the Natural Resources Stewardship Circle (NRSC) with a view to significantly increase their positive impact on the preservation of biodiversity [3]. Cosmetic Valley launched its eco-responsibility charter in 2009, giving a commitment to guarantee the positive integration of industry activities in harmony with biodiversity [4].

Table 5.2 Examples of cosmetic ingredients from natural resources.

- Oils, fats and butters
- Surfactants, emulsifiers and other derivatives of vegetable oils
- Essential oils, absolutes and resinoids from plant material for use in fragrance
- Solvent extracts from plants with active cosmetic properties
- Natural colours
- Clays and other minerals

Source: Adapted from Andrew Jones, Fair Venture Consulting Ltd.

At the international multi-stakeholder level, the United Nations Conference on Trade and Development (UNCTAD) – through its BioTrade Initiative – launched in 2010 the Fashion and Cosmetics BioTrade Platform to bring together the fashion and cosmetics industries to promote collaboration and contribute to on-going international efforts, further sustainable use of biodiversity and build awareness and educate on its importance.

Despite the increased recognition of the importance of biodiversity, the authors of this chapter believe much still remains to be done.

5.1.1 The Critical Loss of Biodiversity and its Impact on the Cosmetics Industry

Along with climate change, it is recognised that the loss of biodiversity is one of the two major challenges faced by the planet today.

Biodiversity is currently being lost at unprecedented rates due to human activities. It is estimated that recent species extinction rates are 100 to 1000 times their pre-human levels; the rate is expected to increase 10 times at existing levels [5].

The Millennium Ecosystems Assessment (MA) found that, since the second half of the twentieth century, humans have changed ecosystems more rapidly and extensively than in any comparable time in human history. Habitat loss and degradation create the biggest pressure on biodiversity worldwide, resulting in substantial and largely irreversible loss of genetic diversity [6].

As the variety of species and ecosystems continues to decline, both the quality and quantity of raw materials and other ecosystem services crucial to the cosmetics industry are becoming threatened. This could translate into increased operational costs for companies because of rising prices of raw materials and restrictions on access to them.

Biodiversity is a driver of innovation in the cosmetics industry. Experts estimate that only around 10% of species are known to man. The remaining 90% could provide a wealth of knowledge and new scientific discoveries for the industry. However, with each species that becomes extinct, the potential discoveries disappear with it.

The continued pressure on biodiversity is likely to result in increased regulation at the international level. Expansion of protected areas where access to natural resources is either restricted or prohibited, or new national and international legislation on access to biological and genetic resources, could bring new challenges. The recently adopted Nagoya Protocol on Access and Benefit Sharing will impact the way the cosmetics industry accesses and carries out research on genetic resources, as well as how it will share the benefits from their use.

Environmental-related trade regulation is also of critical importance for the industry. A number of crucial ingredients are currently regulated under the Convention on International Trade in Endangered Species of Wild Fauna and Flora (CITES).

With changing consumer preferences and the rise of ethical finance, biodiversity will become more relevant for businesses trying to strengthen their reputation, their market positions, as well as their products and their investment base. It is likely that businesses who lag behind on biodiversity practices will see increased risks from these elements, whereas those companies who implement and communicate sustainable practices will be able to capitalise on new markets, new revenue streams and improved access to capital.

It is therefore important that the cosmetics industry better understands its dependence and impact on biodiversity, and that it develops adequate strategies to manage them more effectively.

5.2 WHY SHOULD THE COSMETICS INDUSTRY CARE ABOUT PROTECTING BIODIVERSITY?

In the cosmetics industry, nature is established as a source of fundamental inputs and ingredients. Now, companies are beginning to also recognise the importance of taking steps to conserve biodiversity and to ensure its sustainable use. In part, this trend responds to developments in the legal and policy framework focusing on biodiversity. The need to engage the private sector as a driver of change and good practice is becoming a central point in most international and environmental discussions. For example, discussions on the Green Economy highlight the positive interrelationship between economy, environment and social equity, as well as the role of the private sector in enhancing such dynamics. Calls for business engagement in conservation tend to focus on specific projects or initiatives, but it is increasingly clear that for companies working with natural ingredients, the most significant contribution to biodiversity lies in ensuring ethical sourcing across business practices. If done well, the sourcing of natural ingredients enhances the value of biodiversity at a local level, which can both motivate conservation efforts and promote socio-economic development. The Convention on Biological Diversity (CBD) acknowledges that conservation of biodiversity is only attainable through its sustainable use, and the fair and equitable sharing of the resulting benefits. Other related international instruments, such as the Nagoya Protocol on access and benefit sharing, establish concrete principles for the utilisation of biodiversity to advance equity, conservation and sustainable use. In Section 5.2.2, international discussions on ecosystems services are described; these are also relevant in the business and biodiversity context.

5.2.1 Biodiversity as a Sound Business Strategy

Strategic decisions by companies are also leading them towards ethical sourcing and biodiversity conservation. Cosmetics Europe – The Personal Care Association has recognised that issues linked to natural resources and biodiversity are among the primary challenges for sustainability in the industry [1]. Loss of biodiversity is a real concern for businesses. For instance, companies developing fragrances and flavours are increasingly worried about competing land uses and stress on ecosystems. The shrinking resource base will impact the quantity and quality of available fragrances.

Ethical sourcing involving biodiversity not only recognises the need to secure the source of natural ingredients, but also works to improve the traceability, soundness and sustainability of the relevant supply chains. Ethical sourcing involving biodiversity obliges companies to consider and improve their policies and practices on sourcing, research and development, supply chain management, sales and marketing, compliance and monitoring and evaluation. Companies thus ensure that their own policies, as well as practices along their supply chains, promote the conservation and sustainable use of biodiversity and improve livelihoods at the local level. It avoids problems arising in the future, strengthening partnerships and securing the supply chains.

Biodiversity plays a leading role as the basis of – or inspiration for – many new ingredients and products in the cosmetics industry, thus ethical sourcing means investment in innovation. Leading companies like L'Oréal recognise that 'plant-based ingredients are a huge source of innovation'. [7] Approaches aimed at ensuring sustainable sourcing practices with regards to natural ingredients are also a way to put innovation at the centre of company strategies. Indeed, ethical practices are particularly relevant for research and development, which require adequate protocols to be followed for access to biodiversity and any associated traditional knowledge, and all relevant permits secured. Patent policies need to consider potential questions about the origin and legal access to biological or genetic resources mentioned. Ensuring biodiversity provides continued inspiration is particularly poignant in the context of the continuing rise of consumer interest in natural products.

Natural ingredients are increasingly highlighted by cosmetics brands in their marketing activities. Advertising focuses on the 'naturalness' of products, with key ingredients and their plant sources promoted by brands to attract consumers. Increasing references to biodiversity and nature in product marketing bring higher expectations from consumers. There is a growing interest in where these ingredients come from and how they have been sourced, as well as the preservation of the species and ecosystems involved. Biodiversity can be an extremely powerful message; it is important for companies to communicate their efforts to safeguard biodiversity through ethical practices. Businesses are finding they can effectively educate consumers about biodiversity, while highlighting the unique values and qualities in their products.

5.2.2 Ecosystem Services

The CBD defines an ecosystem as 'a dynamic complex of plant, animal, and micro-organism communities and the non-living environment interacting as a functional unit'. Ecosystems provide human beings with a series of benefits, many of them crucial to their survival. These benefits are called ecosystem services.

The Millennium Ecosystem Assessment describes four types of ecosystem services: provisioning services, which are those that can be extracted from nature such as goods and products; regulating services that moderate natural processes; cultural services, which are non-material benefits that contribute to spiritual and recreational advancement of people; and supporting services that maintain all other services.

Ecosystem services are economically significant for the cosmetics industry. Furthermore, the industry relies heavily on these services and thus should have a direct interest in preserving them. Table 5.3 gives some examples of this reliance.

According to projections of the Millennium Ecosystems Assessment, in the next 40 years provisioning services – especially for crops, livestock and aquaculture – will continue to be enhanced mainly because of the activity of humans and the

Table 5.3 Examples of the cosmetic industry's direct dependence on ecosystem services.

Ecosystem Service	Benefits to the Cosmetics Industry
Provisioning	Natural ingredients Fibres and textiles for packaging Fresh water for production and transformation processes Genetic resources and information to further innovation Biochemicals
Regulating	Pollination Water purification Natural hazard regulation such as floods, drought and land degradation Pest regulation Disease regulation
Cultural	Tourism and other recreational activities requiring personal care products Design inspiration Ethical and existence values that shape consumer aspirations
Supporting	Photosynthesis Nutrient cycling Soil formation

Source: Adapted from WBCSD Ecosystems Training module 1 © WBCSD, 2013.

transformation of habitats. This, however, will likely be at the cost of other provisioning services of vital importance for the cosmetics industry, such as the variability of genetic resources, biochemical, fresh water and wild collected ingredients. In addition, it is expected that supporting, regulating and cultural services will be further degraded [8].

5.3 HOW IS THE POLICY ARENA CHANGING AND WHAT IMPLICATIONS DOES THIS HAVE FOR THE INDUSTRY?

International commitment to halt the loss of biodiversity has grown in recent years. However, these efforts have not been successful. Evidence is the failure of international community to meet the commitments agreed in 2002 to achieve by 2010 a significant reduction in the current rate of biodiversity loss at the global level [9].

As a result, in 2010 governments from 193 countries adopted a landmark agreement setting out a Strategic Plan for Biodiversity 2011–2020. In recognition of the urgent need for action, the United Nations General Assembly also declared 2011–2020 as the United Nations Decade for Biodiversity.

5.3.1 The Convention on Biological Diversity

The CBD was opened for signature at the Earth Summit on 5 June 1992, coming into force on 29 December 1993. It is the first global international legally binding treaty to cover all aspects of the conservation and sustainable use of biological diversity. It recognises in international law that the conservation of biodiversity is a common concern of humankind and is an integral part of the development process. The CBD has three main objectives: the conservation of biodiversity, the sustainable use of its components, and the fair and equitable sharing of benefits arising from its utilisation.

At present, 193 Parties are legally bound to implement its commitments that include: substantive provisions on measures and incentives for the conservation and sustainable use of biodiversity; regulating access to genetic resources and traditional knowledge; equitable and fair sharing of benefits arising from the use of genetic resources; public awareness and education; impact assessment; and access to and transfer of technology. One major drawback of the CBD is that the United States of America is one of three countries that have not yet ratified the convention.

The convention covers the rapidly expanding fields of biotechnology, and research and commercial use linked to genetic resources through its two protocols: the Cartagena Protocol on Biosafety and the Nagoya Protocol on Access to Genetic Resources and the Fair and Equitable Sharing of Benefits Arising from their Utilization, respectively.

The responsibility to implement the CBD lies with the individual countries, which must develop a range of measures and activities in order to put into practice their obligations. In 2012, nearly all countries have adopted strategies and policies to conserve species and habitats. Some 170 countries have formulated national biodiversity strategy plans.

Through the adoption of Decision VIII/17 in 2006 [10], Parties to the CBD officially recognised the important contribution that business can make to the implementation of the three objectives of the convention. This recognition has been growing since; business has been referenced directly in subsequent decisions and work programmes of the convention. The COP 10 decided to involve the business community in the preparation of the National Biodiversity Strategy and Action Plan (NBSAP), creating a new opportunity for business to get involved in national policy making.

5.3.2 The Strategic Plan for Biodiversity

The tenth meeting of the Conference of the Parties of the CBD (COP10), in October 2010, in Nagoya, Japan, adopted a revised and updated Strategic Plan for Biodiversity. The Strategic Plan comprises a global vision, a mission, strategic goals, and a new set of 20 targets, collectively known as the Aichi Targets [11]. Box 5.1 provides an overview of the Strategic Plan.

Box 5.1 The Strategic Plan at a Glance

The Strategic Plan lays out a vision of a world of 'Living in harmony with nature' where 'By 2050, biodiversity is valued, conserved, restored and wisely used, maintaining ecosystem services, sustaining a healthy planet and delivering benefits essential for all people'.

Its mission is to take effective and urgent action to halt the loss of biodiversity in order to ensure that, by 2020, ecosystems are resilient and continue to provide essential services, thereby securing the planet's variety of life, and contributing to human well-being and poverty eradication.

The Strategic Plan includes 20 headline targets for 2015 or 2020 organised under five strategic goals:

Strategic goal A – Address the underlying causes of biodiversity loss by mainstreaming biodiversity across government and society

- Target 1: Public awareness increased
- Target 2: Values of biodiversity recognized

- Target 3: Incentives reformed
- Target 4: Sustainable production and consumption promoted.

Strategic goal B. Reduce the direct pressures on biodiversity and promote sustainable use.

- Target 5: Habitat loss reduced
- Target 6: Towards sustainable management of fisheries
- Target 7: Sustainable management (agriculture, aquaculture and forestry)
- Target 8: Pollution reduced
- Target 9: Invasive alien species combated
- Target 10: Pressures on vulnerable ecosystems impacted by climate change or ocean acidification minimized.

Strategic goal C: To improve the status of biodiversity by safeguarding ecosystems, species and genes

- Target 11: Protected areas increased
- Target 12: Extinction prevented
- Target 13: Genetic diversity maintained.

Strategic goal D: Enhance the benefits to all from biodiversity and ecosystem services.

- Target 14: Ecosystems are restored and safeguarded
- Target 15: Ecosystem resilience enhanced
- Target 16: Nagoya Protocol is in force.

Strategic goal E. Enhance implementation through participatory planning, knowledge management and capacity-building

- Target 17: National biodiversity strategy and action plan developed
- Target 18: Traditional knowledge respected
- Target 19: Knowledge and technologies improved and shared
- Target 20: Financial resources increased.

Source: CBD Decision X/2

The cosmetics industry has a unique opportunity to go beyond mitigating negative impacts and become an important driver in ensuring the conservation of biodiversity and contributing to the achievement of the Aichi Targets.

For example, the cosmetics industry can further enhance its knowledge on the sustainable use and management of species that provide it raw materials, as well as on the development and application of greener production practices, and thus contribute to targets 7, 8, 12 and 19.

Since consumers look at the appearances of cosmetic products and respond to marketing, brands can communicate sustainability elements via their marketing activities. Thus, biodiversity has the potential to filter down into the broader population by the marketing of cosmetic products, contributing to targets 1, 2 and 4 of the strategic plan.

The use of traditional knowledge for commercial research and development by the industry could become an entry point for positive contribution to targets 13, 18 and 19.

As mentioned earlier, businesses have been formally invited to take part in the formulation of the NBSAPs which would entail a direct contribution to meeting target 17.

These are just a few examples of the positive role that the cosmetics industry can play in contributing to the global efforts to halt the loss of biodiversity by 2020. The cosmetics industry should grasp this opportunity because, as mentioned in the previous sections, its long-term success depends on maintaining biodiversity and healthy ecosystems.

Apart from the aforementioned benefits, encouraging biodiversity can advance broader sustainable development objectives, such as poverty reduction, job creation and gender empowerment.

5.3.3 The Nagoya Protocol

Another achievement of COP10 was the adoption of the Nagoya Protocol on Access to Genetic Resources and the Fair and Equitable Sharing of Benefits Arising from their Utilization.

Due to repeated cases of 'biopiracy' and the limited flow of benefits towards developing countries that provided genetic resources, countries set forth the development of a more specific and effective international regime. The seventh meeting of the Conference of the Parties to the CBD, held in Kuala Lumpur in 2004, adopted the mandate to initiate negotiations which concluded in Nagoya, Japan in October 2010 [12].

The Nagoya Protocol will have a direct impact on industries that rely on biodiversity to source ingredients. Thus, the cosmetics industry will need to understand and comply with the requirements and obligations of the protocol.

The main aim of the protocol is to effectively implement Article 15 of the CBD: dealing with access to genetic resources and requiring contracting parties to establish 'appropriate, effective and proportionate' measures to ensure that genetic resources and traditional knowledge used within their jurisdiction have been accessed on the basis of prior informed consent and mutually agreed terms, as required by the country of origin [13].

The protocol defines three concepts – 'utilization', 'biotechnology' and 'derivative' – that are key to understanding the implications for the cosmetics industry. A recent study by UNCTAD (X UNCTAD ABS])concludes that an interpretation of the treaty language guided by the Convention on the Law of the Treaties 'seems to leave little doubt that: (a) the "utilization" of genetic resources includes research and development on the biochemical composition of such resources as well as on the individual compounds (derivatives) contained therein; (b) the commercial exploitation of the outcomes of such research and development is subject to the protocol's benefit sharing provisions; and (c) access to genetic resources with the purpose of exploiting the commercial value of derivatives is also subject to the protocol'.

Research on the properties of extracts and molecules from plants, for example, and their development and commercialisation as ingredients in cosmetics, would thus now be distinctly subject to access and benefit sharing and prior informed consent requirements [14].

In this regard, while many of the activities related to the cosmetics industry will remain outside the scope of the Nagoya Protocol, its impact will be particularly important for research and development on genetic resources aimed at identifying and commercially exploiting new biochemical compounds of interest. These compounds are the basis for a wide range of cosmetic ingredients.

On a positive note, it is likely that the protocol will provide an improved legislative structure that will promote more transparent regulations for the cosmetics industry.

5.3.4 The Convention on International Trade of Endangered Species of Wild Flora and Fauna (CITES)

CITES came into force in 1975, with the aim to ensure that international trade was not so excessive that it threatened the survival of wild animals and plants. Today, it has 175 nations as Contracting Parties, providing varying degrees of protection to more than 35 000 species of animals and plants.

Many of the iconic species used to symbolise CITES – such as the tiger, the panda bear, whales and elephants – are endangered. Although the name may suggest that CITES only regulates trade of endangered species, these actually represent only about 3% of the total number of species under the convention.

In this regard, CITES focuses its efforts to ensure the sustainability of the trade in order to safeguard biodiversity for the future through three main elements:

- Identify species currently endangered by trade (Appendix I) or vulnerable to reaching that state (Appendix II).
- To stop trade where it is known or suspected to cause extinction (Appendix I).

- To use a permit system for trade in Appendix II species that essentially certifies that trade is legal and sustainable.

Species that are considered severely depleted by trade and at risk of extinction (which total approximately 1000 species) are listed in Appendix I of the convention. An international trade ban is imposed on them with the aim to foster a population recovery.

Species that are identified as likely to become endangered by trade, unless the extent of wild harvest is regulated and becomes no longer detrimental to their survival, are listed in Appendix II. This Appendix accounts for more than 90% of the total of species listed in the convention, approximately 34 800.

All international trade in Appendix II species, parts (including for example roots, seeds, barks, leafs), intermediate inputs (oils, extracts, waxes, etc.) or products made from them (lipsticks, creams, body washes, etc.) need a CITES export permit, issued by the exporting state. Many states require the presentation of a reciprocal CITES import permit, however this is considered part of stricter domestic measures.

Species not themselves threatened by trade can be listed in Appendix II if they are considered like a threatened species; they cannot be easily identified and distinguished from CITES-listed species by border control officials.

Appendix III, the smallest with approximately 270 species, was created with the objective of allowing individual countries to put species they felt needed to be regulated without going through the normal procedure. As such, this appendix contains species that are protected in at least one country that has asked other CITES parties for assistance in controlling the trade.

In a survey carried out by UNCTAD in 2009 on about 50 cosmetic companies, it was found that above all other policies and legislation, CITES legislation was considered as the one that can have the most significant impact on a company's business. This is not surprising since many wild collected plant species are used in the industry. The well-known examples of *euphorbia antysiphylica*, *bulnesia sarmiento* and *aniba rosaeodora* confirm this; the last two have been added to Appendix II recently. Table 5.4 gives a sample list of cosmetic ingredients subject to CITES legislation.

The UNCTAD 2009 survey also shed light on the difficulties companies have in dealing with species listed under Appendix II. Some companies decided not to use these ingredients at all, which can have a negative impact on suppliers. Other companies recognise that the CITES controls permit the sustainable trade of biodiversity under certain conditions and the possibility to sustainably use a greater range of ingredients.

In recent years, the cosmetics industry has increasingly recognised the opportunities of closer participation in CITES consultation and decision-making processes. One successful outcome of this closer participation is the clarification of a provision of the CITES Convention by a resolution agreed at the last conference of the parties,

Table 5.4 Examples of cosmetic ingredients subject to CITES.

Ingredient
Aglaia odorata flower oil
Aloe ferox leaf extract
Aniba rosaeodora wood oil
Aquilaria agallocha stem extract
Bulnesia sarmientoi extract
CACTACEAE Spp
Caesalpinia echinata wood extract
Cedrela odorata wood oil
Cyathea cumingii leaf extract
Cycas revoluta leaf powder
Dionaea muscipula extract
EUPHORBIA (various incl. Candelilla)
Gastrodia elata root extract
Guaiacum officinale wood oil
Hydrastis canadensis extract
Nardostachys jatamansi oil
Paphiopedilum maudiae flower extract
Prunus africana bark extract
Pterocarpus santalinus wood extract
Taxus cuspidata leaf extract

Source: CITES

held in Doha in March 2010, where exemptions from requiring CITES export and import permits were nominally agreed for personal care products.

5.4 BIODIVERSITY BAROMETER: CONSUMER VIEWS AND EXPECTATIONS ON BIODIVERSITY

In Section 5.2, reasons were given why cosmetic companies should be concerned with protecting biodiversity. One reason is consumer expectations; this is particularly important in a consumer-facing industry, such as cosmetics. If consumers are aware of biodiversity and its importance, they are likely to demand more from companies in their use of it.

While biodiversity has been at the heart of many recent political discussions and is on the agenda of numerous international environmental organisations, it does not necessarily follow that the average consumer is aware of its importance. Since 2009, the Union for Ethical BioTrade (UEBT) has been conducting an annual study with the aim of gauging how much consumers know about biodiversity and what their expectations are towards companies that source from it. This study, called the Biodiversity Barometer, has now been performed in some countries over a number of years. The results are encouraging.

5.4.1 Biodiversity Awareness is Growing

The UEBT Biodiversity Barometer was initially performed in France, Germany, the United Kingdom and the United States. Online interviews were performed with 1000 consumers in each country to ask them whether they had heard of biodiversity and if so, could they define it. In 2009, 56% of respondents in these four countries said they had heard of biodiversity. In subsequent years, new countries have been added to the UEBT Biodiversity Barometer; in 2012, eight countries were surveyed (Brazil, France, Germany, India, Peru, Switzerland, the UK and the US). In these countries, on average 63% of consumers said they had heard of the term. Some countries had particularly high consumer awareness of biodiversity: Brazil (97%), France (95%) and Switzerland (83%). These countries also showed a significant increase in awareness over the last few years.

Having heard of the term biodiversity is not necessarily the same as understanding what it actually means. Awareness of biodiversity is generally high, although significant differences exist between countries, even within the same region. The understanding of biodiversity, measured through the number of people that provided correct definitions of biodiversity, is often very limited: In 2012, in no country did it exceed 50%. This low understanding demonstrates the need for more education on biodiversity, but also the potential for further understanding. Table 5.5 shows that a number of the definitions given by consumers were close to being correct.

Table 5.5 Cosmetic companies reporting about sustainability and biodiversity.

What are companies saying about biodiversity?	2009	2010	2011	2012	Variation 2012 vs 2009
Companies reporting on sustainability development	44%	52%	52%	54%	⇧ + 10
Companies reporting biodiversity	13%	21%	27%	31%	⇧ + 18
Companies reporting on biodiversity sourcing practices	9%	12%	19%	21%	⇧ + 12
Companies mentioning biodiversity related issues like traditional knowledge and intellectual property rights	2%	3%	5%	4%	⇧ + 2

Basis : UEBT analysis of top 100 beauty companies (WWD)

Source: © UEBT.

5.4.2 Increased Awareness Brings Greater Expectations

As consumers become more informed, their expectations of the private sector become higher. Consumers want to know more about how their products are made, who makes them, where they are made and where the ingredients are coming from. This has been evident in many industries. According to results from the 2012 Biodiversity Barometer, 85% of consumers look for natural ingredients in cosmetics products and 69% pay attention to where the natural ingredients come from. In emerging markets, such as Brazil, India and Peru, the numbers are even higher: 86% of consumers in these countries look for natural ingredients and 80% consider where these ingredients come from. On average, 75% of consumers consider that the private sector has an important role to play in the conservation and sustainable use of biodiversity.

Consumers would like to know more about how a company sources its natural ingredients. Not only did consumers say they want to know more about companies' sourcing practices, they also said they would be willing to act if a company was not behaving in an ethical way. In 2012, over 80% of respondents stated they would boycott a brand if its biodiversity sourcing was not environmentally and socially responsible. In Brazil, where consumers are particularly aware of their country's rich natural heritage, 88% of respondents said they would boycott a brand if it was not engaging in ethical sourcing. Of course, there is always a difference between what consumers say they will do and what they actually do, but such figures suggest brands have a lot to lose if their practices do not fit consumers' expectations.

5.4.3 Opportunities for Pioneering Companies

These results suggest that consumers overwhelmingly want to know more about how companies source ingredients. Embracing transparency and communicating fully on biodiversity and sourcing practices can help companies stand out from their competitors. Yet corporate communication on biodiversity in the cosmetics and personal care industry is still not common practice. Only a small number of the world's top cosmetic and personal care companies (UEBT analysis based on WWD top 100 cosmetic companies) communicate on biodiversity: in 2012, it was only 31% of the top companies (see Figure 5.1). Yet this number is significantly greater than just a few years ago. In 2009, only 13% of the top cosmetics companies mentioned biodiversity in their corporate communications. Clearly, as consumer awareness of biodiversity continues to grow, so does the response of cosmetic companies.

The messages used when communicating on biodiversity are also important. In 2011, the Biodiversity Barometer asked consumers what would make them purchase a cosmetics product containing natural ingredients sourced from Africa.

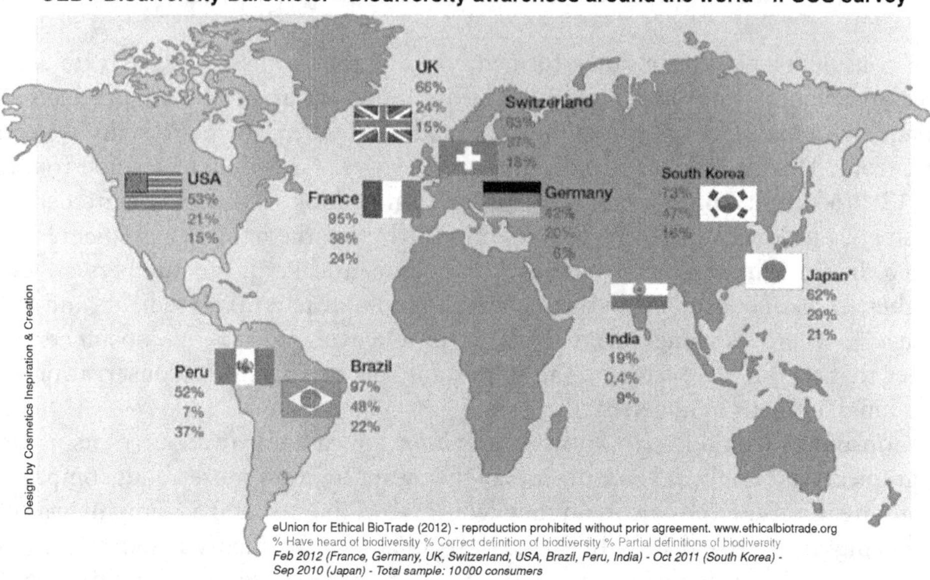

Figure 5.1 Awareness of biodiversity, from UEBT Biodiversity Barometer (2012). *Source*: © UEBT.

Consumers responded that protecting local biodiversity and improving livelihoods of the local population were motivating factors when buying such products. Focusing on these two characteristics will help create strong messages to engage consumers.

The results of the Biodiversity Barometer suggest an increasingly ethical and demanding consumer looking for more than just a cosmetics product. Today's consumer looks for products that can provide multiple benefits, including environmental or ethical. Providing consumers with a way to contribute to a more sustainable world is a great way of forging lasting relationships. By focusing on ethical and sustainable use of biodiversity, brands can offer consumers a way to make a difference, thereby helping create a loyal following.

Other opportunities also exist for companies that want to expand their sustainability portfolio beyond ethical sourcing practices. Two examples are outlined below.

5.4.3.1 Payment for Ecosystem Services: a Biodiversity/Climate Change Opportunity

The Economics of Ecosystems and Biodiversity (TEEB) for policy makers [14] define Payment for Ecosystem Services as voluntary transaction where a well-defined ecosystem service is 'bought' by at least one ecosystem service buyer from at least one ecosystem service provider, if and only if the ecosystem service provider secures the ecosystem service provision. In broad terms, Payments for Ecosystem Services refer to the practice of a beneficiary paying landowners in exchange for managing their land to provide some sort of ecological service. These programmes promote the conservation of natural resources in the marketplace.

In recent years, payments for Reducing Emissions from Deforestation and Forest Degradation (REDD) has become more relevant for both policy makers and business leaders. The cosmetics industry could further benefit from the inclusion of conservation, sustainable management of forests and enhancement of forest carbon stocks (REDD+) where new financial incentives could be seized through biodiversity policies, such as the strengthening of the value chain of products that create livelihood opportunities in forest areas and reduce deforestation.

5.4.3.2 Offsets

Offsetting refers to measurable conservation outcomes resulting from actions designed to compensate for significant residual adverse biodiversity impacts and persisting after appropriate prevention and mitigation measures have been implemented [15].

The goal of biodiversity offsets is to achieve no net loss of biodiversity on the ground with respect to species composition, habitat structure and ecosystem services.

The biodiversity offsetting practices vary from project to project, with the implementation taking place either within the affected habitats or outside.

In general terms, offsetting can take place in three different ways:

1. The company or direct partner (such as an NGO) implements the offset.
2. The company pays the government to implement the offset.
3. The company buys 'credits' from the landowner of conservation bank sufficient to offset its impacts.

In this context, a company that has residual adverse impacts – for instance of water pollution that cannot be further reduced or mitigated – can invest in conservation activities of other habitats in a way to achieve no net loss.

The uses of 'offsets' are controversial as the offsets do not provide a solution for the original impact, and they might create perverse incentives that enable companies to continue degrading ecosystems. The offsets may not adequately compensate for the value of biodiversity lost and the protection of critical species and ecosystems; they also may not compensate for outcomes that would otherwise not occur.

For this reason, it is important that offset projects are designed and implemented with the participation of all relevant actors, and within a clear regulatory and legal framework that allows for the adequate valuation, measurement and optimisation of conservation impacts.

5.5 ETHICAL SOURCING IN PRACTICE

Previously in this chapter, the importance of biodiversity for the cosmetics industry was discussed and why companies should make efforts to promote its protection. In this final section, it will be shown how companies can have a positive impact on biodiversity through their sourcing practices.

5.5.1 Putting Ethical Sourcing of Biodiversity into Practice

At the international level, treaties such as the Convention on Biological Diversity (CBD), which has the support of over 190 countries, establish common objectives and principles for the conservation and sustainable use of biodiversity. Yet it is not always easy for businesses, particularly small and medium-sized enterprises, to determine what these principles mean on the ground. There are questions as to which issues are most relevant as companies develop, produce and market natural ingredients, or what measures need to be taken at what level. There is a need for methodologies, as well as systematic guidance and monitoring of company practices.

In this context, Ethical BioTrade has evolved a framework for the implementation of ethical sourcing practices, which has become a reference for examples of good practice and successful approaches. The Ethical BioTrade standard, developed in 2007 through a global consultation process, and managed by the Union for Ethical BioTrade (UEBT), is based on the objectives and principles of the CBD [16]. Through a series of principles, criteria and indicators, it provides detailed orientation on how to put ethical sourcing of biodiversity into practice. For example, the Ethical BioTrade standard explains how organisations can ensure sustainable use of biodiversity, including training those involved in sourcing activities on good collection practices. As described in Figure 5.2, the Ethical BioTrade standard also deals with conservation, equitable sharing of benefits, issues linked to productive and financial management, compliance with legislation and clarity on land tenure and use rights.

By following the Ethical BioTrade standard, UEBT members are working to put in place a biodiversity management system that ensures that their sourcing practices recognise and enhance the value of biodiversity at the local level, thereby motivating conservation and improving local livelihoods. Through these efforts, as illustrated

Ethical BioTrade Principles and Criteria

Biodiversity conservation	Preserving ecosystems and promoting practices that contribute to conservation plans and strategies.
Sustainable use	Sourcing practices promote the long-term ability of an ecosystem to function.
Fair and equitable benefit sharing	Use of biodiversity and traditional knowledge occurs on the basis of prior informed consent and the equitable sharing of monetary and non-monetary benefits.
Socio-economic sustainability	Adopting quality management and financial practices that are sustainable and socially acceptable.
Legal compliance	Respecting all relevant international, national and local regulations.
Respect for the rights of actors	Taking into account human rights and working conditions of indigenous and local communities.
Clarity about land tenure	Respecting land tenure and rights over natural resources.

Figure 5.2 Ethical BioTrade Principles.
Source: Union for Ethical BioTrade (© UEBT).

below, UEBT members are actively contributing towards the objectives of CBD, including conservation and sustainable use of biodiversity, and fair and equitable sharing of the benefits derived from its utilisation.

Box 5.2 Why Ethical BioTrade?

A range of factors encourage companies to engage in ethical sourcing practices. For example, Symrise Brazil, a fragrance house with its headquarters in Germany, recently joined UEBT, selecting Ethical BioTrade as a fundamental approach to support its work towards ethical sourcing of natural ingredients. For Symrise Brazil, the choice of Ethical BioTrade is linked to the importance given to biodiversity in its portfolio and strategy – Symrise Brazil recognises that respect for biodiversity, local communities and partners all along the chain is the only way to ensure the long-term supply of the raw materials utilised in its innovations and creations. Moreover, Symrise Brazil, as a business-to-business company, was also responding to customers' – and, ultimately, consumers' – growing demand for ethical sourcing practices. UEBT was selected, because of its standing in the cosmetics industry, expertise in the integration of biodiversity in company management system, independent verification system and technical advice and support on access and benefit sharing.

Box 5.3 A System Approach to Biodiversity

In Ethical BioTrade, biodiversity is more than a line of products or a group of projects; it is a cross-cutting consideration in all company activities. UEBT members must have in place a biodiversity management system. Just like a Quality Assurance system or a Good Manufacturing Practice (GMP) system, this is a set of procedures that seek to ensure that sourcing practices advance conservation and sustainable use of biodiversity, as well as the equitable sharing of benefits. For example, Laboratoires Expanscience, a French cosmetics and pharmaceutical company, has been strongly committed to sustainable development from its inception.

In 2011, Laboratoires Expanscience joined UEBT as part of its efforts to strengthen this commitment and integrate sustainable development goals in its entire business through developing internal policies on biodiversity, as well as promoting ethical practices along its supply chains. Work is ongoing to enhance the way in which the internal quality assurance system and purchasing department of Laboratoires Expanscience take into account biodiversity issues. At the supply chain level, Laboratoires Expanscience is formalising its own responsible sourcing reference system, including specific commitments and good practices to ensure the security of its supplies and a positive impact on local development. The research and development department is working to promote and monitor the adoption of these practices throughout the supply chains.

5.5.2 Conservation of Biodiversity

With the loss of biodiversity continuing unabated, conservation remains a priority among international sustainable development objectives. Conservation of biodiversity is also Principle 1 in the Ethical BioTrade standard, which recognises the need for sourcing practices to support the preservation and restoration of species, habitats and ecosystems. Here, the focus is not on how the natural ingredients are collected or harvested – this is dealt in Principle 2 on sustainable use – but rather on the link between sourcing activities and the integrity of the ecosystem in which they take place. For example, is there illegal logging taking place around the sourcing area? If so, what can be done to help prevent such logging? The Ethical BioTrade standard requires companies to identify possible threats and to contribute to addressing them. In addition, sourcing activities should neither involve conversion of pristine ecosystems, nor introduce invasive alien species or genetically modified organisms. Sourcing activities should also align themselves with traditional practices and local strategies that contribute to preserving biodiversity in the area.

Box 5.4 Conserving Forests in Madagascar

UEBT members Aroma Forest and Man and the Environment (MATE) are working together to conserve biodiversity in some of the last remaining forests in the Eastern coast of Madagascar. Local communities in these forests are extremely poor, in spite of the wealth and potential of biodiversity around them. MATE is a non-profit organisation focusing on advancing biodiversity conservation by involving local people in need. Aroma Forest is a company producing high-quality essential oils for some of the top cosmetic and fragrance houses. Through their collaboration, MATE and Aroma Forest aim to show that the link between private sector and local communities, if conducted on the basis of ethical sourcing practices, can improve livelihoods of local communities and constitute a driver for biodiversity conservation.

For example, Aroma Forest actively contributes to rehabilitating the ecosystem in which it is working, which has already been significantly degraded. Aroma Forest, through MATE, is providing community associations with seedlings for reforestation purposes. There are also discussions on building a community-run nursery. Aroma Forest and MATE also monitor any potentially negative effects of their activities and work to mitigate them. For instance, communities are planting deep-rooted grasses to help protect against erosion and desertification in areas around paths and distilleries. Treatment of waste coming from the distilleries is also being improved, including through a programme for the reuse of solid waste as fuel. In this way, Aroma Forest is confident that its engagement will have an overall positive effect on the biodiversity.

Box 5.5 Preserving the Amazon Rainforest

One in ten known species in the world lives in the Amazon Rainforest, with many more waiting to be catalogued or discovered. For companies working in and around the Amazon, conservation of biodiversity is not only an ethical imperative, but also an investment in the sustainable supply of innovative and remarkable ingredients. Through ethical sourcing practices, several UEBT members are actively seeking solutions for conservation and sustainable development in one of the world's most important and threatened ecosystems:

- Candela Peru sources Brazil nut from indigenous and local communities in Madre de Dios, in the Peruvian Amazon. Through its ethical sourcing practices, the company is improving local livelihoods and promoting the sound management of forest resources. Brazil nuts, which must be gathered following established good practices, provide an important income for local communities, as well as promoting the sustainable use of the rainforest.

- Beraca HPC (Health and Personal Care division) sources natural ingredients for the food and cosmetics sectors around Belém, in Brazil. It has an internationally recognised programme for the valorisation of biodiversity, which constitutes a source of strategic and commercial opportunities, as well as an approach to supporting regional development and biodiversity conservation. Beraca HPC is also looking to integrate company know-how on water and sanitation to improve the quality of life for people in the Amazon region.
- In 2011, Natura Cosmeticos SA, one of the leading cosmetic companies in the world, launched its Amazon Program. The aim of the Amazon Program is to contribute to the sustainable development of the Amazon through science, technology, innovation and high-value supply chains in the region. Its approach features joint efforts with other companies, which will act as a network to exchange energy, materials and information – in order to fully reveal the business potential of sustainable use of biodiversity in the Amazon.

5.5.3 Sustainable Use of Biodiversity

In the sourcing of biodiversity, there is an obvious need to ensure that the way in which biological resources are used does not lead to their long-term decline. Principle 2 of the Ethical BioTrade standard focuses on the sustainable use of biodiversity, requiring sourcing activities to be based on management documents that consider harvest, productivity and regeneration rates. For instance, collection and cultivation practices should not be negatively affecting the population of the sourced species. In practice, this means UEBT members must train workers involved in collection and/or cultivation in good practices, and make sure that cultivation and harvesting are performed in a way that will allow similar yields in the future. Moreover, a monitoring system should be in place to allow for continual adjustment of these practices: what is known as adaptive management. Principle 2 also addresses the need for organisations to implement appropriate mechanisms to prevent or mitigate negative environmental impact. Such mechanisms should deal with issues such as use of agrochemicals, protection of water resources, preservation of soil and air quality, and waste disposal.

Box 5.6 Securing Sustainable Supply of Ratanhia

Ratanhia is a traditional medicinal plant of the Peruvian Andes, as well as a natural ingredient used in oral and dental care products. As its popularity in local and export markets surged, Ratanhia stocks plummeted. It became imperative to ensure the sustainable harvest of Ratanhia and the protection of its habitat.

In this context, and following its commitment to respecting nature and caring for the rights of future generations, Weleda AG, a manufacturer of natural cosmetics and medicines, has been implementing sustainable collection practices for Ratanhia for over 40 years. Through collaboration with local farmers and the Peruvian environmental authorities, Ratanhia is now protected on 5000 acres of certified organic land. Collectors are trained in good practices: for every plant that is extracted from the earth, five Ratanhia seeds are replanted in the same location. Through this project, Weleda has secured the supply of Ratanhia for use in its oral care formulations and ensured a sustainable life for the Ratanhia plant.

Box 5.7 Sustainable Use in Madagascar

Serdex, a division of Bayer Santé Familiale and UEBT member since 2009, produces plant extracts for pharmaceutical and cosmetic applications. Its business has traditionally focused on *Centella asiatica*, a plant sourced sustainably from Madagascar. For example, collection areas and quantities for *Centella asiatica* are always established on the basis of relevant permits, and clearly defined and transmitted to local collectors. Continuous training is provided to collectors and harvesters to ensure good collection practices, and field visits are conducted regularly to monitor the application of these practices. Good collection practices are also controlled through the quality of the dried plants. In this way, Serdex can assure a top-quality product and realise its commitment to Ethical BioTrade and sustainable development.

Box 5.8 Natura and Its Commitment to Organic Alcohol

In 2006, as part of its growing commitment to the sustainable use of biodiversity, Natura Cosmeticos SA, began evaluating possibilities for sourcing commodity ingredients from sustainable supply chains. In 2007, Natura signed a partnership with Native, an organic food enterprise, to start sourcing its alcohol from organic agricultural sources. It was an important initiative that has added value to Natura products that use alcohol, which is now 100% organic and reflects related good production practices. The move has had a significant impact. According to Kantar World Panel Data from 2011, the fragrance market represents 62% of the total Brazilian cosmetic market, whilst Natura has 32% share of this market.

5.5.4 Fair and Equitable Benefit Sharing

Equitable sharing of benefits derived from the use of biodiversity is at the core of Ethical BioTrade, seeking to ensure that sourcing activities contribute to local sustainable development goals, provide incentives at the local level for the conservation of biodiversity and recognise the contributions of local actors. Principle 3 of the Ethical BioTrade standard deals with the fair and equitable sharing of benefits derived from the use of biodiversity, both in the sourcing of natural ingredients and, more specifically, in research and development based on biological resources and associated traditional knowledge.

A cross-cutting consideration for fair and equitable sharing of benefits ensures that discussions along the supply chain, and particularly at the local level, are transparent and built on dialogue and trust. This requires negotiations with all relevant stakeholders, which in some cases extend beyond the organisations along the supply chain, and include other communities, groups or individuals with rights over the resources. Information used in these negotiations must be complete and allow those involved to gain an understanding of the issues and potential impacts at stake. There are also requirements on empowering local actors to actively participate in negotiations, as well as on the documentation of their outcome. For example, as described in Box 5.2, some UEBT members are exploring how biocultural community protocols may be useful towards supporting indigenous and local communities in their engagement in these negotiations.

Box 5.9 A Biocultural Dialogue in Peru

Candela Peru, a UEBT founding member, is working jointly with Asociacion Forestal Indigena (AFIMAD), an indigenous forestry association in the Peruvian Amazon, in natural products that promote sustainable development and improve livelihoods for indigenous communities. Candela Peru and AFIMAD are using the Ethical BioTrade standard as a tool to enhance their partnership. In particular, with the support of UEBT and other partners, Candela Peru and

AFIMAD are exploring tools to promote mutual understanding in their negotiations. For example, in 2011, a methodology based on experiences with biocultural community protocols was utilised to support the AFIMAD communities in clarifying social, cultural and environmental expectations or aspects of their relationship with Candela Peru, promote respect their biocultural heritage and advance work towards Ethical BioTrade practices. Indeed, for Candela Peru, this was an opportunity not only to learn more about its partners, but also to explain its own values and approaches to working with biodiversity, as defined by its membership in UEBT.

The outcome of the dialogue between AFIMAD and Candela Peru was the establishment of an 'Agreement of Principles and Commitments' between both parties. This document includes the principles on which their working relationship is based, the specific commitments of each party to advancing work under the Ethical BioTrade framework; and the concrete next steps for follow up in the context of exploring future projects. The 'Agreement of Principles and Commitments' is an internal document, held by both the company and the communities. Nevertheless, the document will become a reference in the development and assessment of continuing efforts of Candela Peru, as a UEBT member, to implement the Ethical BioTrade standard.

Ethical sourcing practices require companies to pay equitable prices for sourced species and ingredients. Indeed, the income from the collection or cultivation of natural ingredients can and should significantly improve local livelihoods. For example, prices must cover the costs of production, including those linked to conservation, sustainable use and other requirements in the Ethical BioTrade standard. In addition, ethical sourcing practices should be structured and conducted in a way that advances local sustainable development goals, as defined by the producers and their communities. Ethical BioTrade requires companies' policies and practices to address issues such as generating employment, enhancing value added at the local level, establishing long-term partnerships with producers, and working to build their institutional, natural resource management, technical and commercial capacities – as exemplified in the case studies described in Boxes 5.10 and 5.11.

Box 5.10 Supporting Local Development in Colombia

Ecoflora is a Colombian company producing ingredients from local biodiversity for the food, cosmetics and household cleaning sectors. It was founded in 1998 on the basis of strong ethical and environmental principles; it has been a UEBT member since 2009. In 2010, Ecoflora, in partnership with the local non-profit

FUNLAU, was awarded one of the UEBT Community Trading Grants, to commercialise laurel wax, sourced from the fruits of *Morella pubescens*. The aim of the Community Trading Grants is to accelerate the implementation of the Ethical BioTrade Principles with communities; the supply chain will be developed taking into these into account.

In the project on laurel wax, one of the main goals of activities is generating and enhancing the flow of economic, social and environmental benefits from biodiversity – particularly laurel wax – to local communities. These communities have long produced laurel wax as an ingredient for moulds for 'panela', a local delicacy made from sugarcane. Yet the laurel wax produced is generally low quality, sold only in local markets and considered a marginal and non-profitable product. By promoting better extractive procedures, enhancing quality and developing new uses and markets for laurel wax, the project aims to increase demand and enable producers to have more value addition and revenues at the local level. Ecoflora and FUNLAU have an agreement in place committing Ecoflora to buy laurel wax from the producers of the region, through FUNLAU. FUNLAU, in turn, commits itself to ensuring the benefits obtained are distributed equitably among the producers. There are also non-monetary benefits derived from organisation support, training, and technology transfer.

Box 5.11 Creating Opportunities for Women in Swaziland

Swazi Indigenous Products (SIP) was formed in 2004 with the aim of creating new income-generating activities for rural Swazi women. SIP, which is a member of UEBT, works with many local species, wild harvesting seeds for their oil content. The oils are then supplied to the global cosmetics market, as well as used in an own-brand, locally made, finished cosmetic products range.

One of the major species SIP works with is the marula tree (*Sclerocarya birrea*), which is widespread throughout Southern Africa, growing in arid, marginal lands. Swazi women traditionally collect the marula fruits to make beer. This home brew has important cultural significance, as well as providing essential supplementary income for rural Swazi women. The work of SIP has added an extra income-generating step to this traditional activity. Once the fruit has been used, the hard-shelled nut is dried in the sun and can be cracked open to free the oil-containing seeds. Instead of throwing the nuts away after making the beer, the women crack them and supply the seeds for oil.

The company is one 100% community-owned. All profits reach the rural women themselves. The board of directors, half of which are rural women, review prices annually, which means that suppliers are heavily involved in pricing. Prices paid by SIP for the marula oil are some of the highest in the region. SIP is also expanding its range of organic products, which command higher prices and allow a premium to be given to suppliers. Furthermore, SIP has developed a range of locally made finished products to generate additional benefits for the local community, as well as to balance supply and demand. The domestic processing of the high quality oil into finished formulations also enables higher levels of skills development, as well as higher local earning potential. The Swazi Secrets range is currently available in speciality shops in 16 European countries, as well as in the United States.

Biodiversity is an important source of innovation for companies in the cosmetics industry. The CBD establishes a set of principles that guide research and development on the genetic or biological composition of biological resources, seeking to ensure the rights of countries and communities to biodiversity, and associated traditional knowledge, are respected and that their contributions are justly compensated. Companies working with natural ingredients and committed to ethical sourcing practices need to be aware of the principles that guide access to genetic resources and the sharing of benefits resulting from related research, application or commercialisation (ABS). They should also have information on legal or regulatory requirements applicable to its activities and take steps to meeting these requirements.

Box 5.12 Access and Benefit Sharing in Brazil

Access and benefit sharing or 'ABS refers to the set of rules and principles governing the use of genetic resources and associated traditional knowledge, established by the Convention on Biological Diversity (CBD). ABS aims to respect the rights of countries and communities over genetic resources and associated traditional knowledge, as well as to ensure there is fair and equitable sharing of the benefits resulting from the use of these resources. The Nagoya Protocol, a supplementary agreement to the CBD, also establishes rules that aim to put ABS into practice. These international rules are implemented, at the national level, through legislation and regulatory requirements.

Brazil has been at the forefront of efforts to implement ABS, but it is still working to develop clear, comprehensive and practical legislation on the topic. Provisional Measure 2.186–16 regulates access and transfer of components of the genetic heritage and associated traditional knowledge in Brazil, as well as the sharing of benefits from the economic utilisation of products or processes resulting from the access. It creates the Genetic Heritage Management Council (CGEN) as an entity with technical, legislative and regulatory functions on ABS.

Natura Cosmeticos SA, one of the founding members of UEBT and one of the leading cosmetics brands in the world, has actively engaged in the development and implementation of ABS rules in Brazil. Natura is responsible for 68% of the applications received by the Genetic Heritage Management Council (CGEN), the entity responsible for granting access to Brazilian genetic heritage and associated traditional knowledge. To complement and address perceived gaps in the Brazilian legal framework, Natura has also adopted an access and benefit sharing policy, which seeks to put into practice CBD principles and advance compliance with the Brazilian legislation. Moreover, Natura is also recognised as a constructive partner in ongoing discussions on the revision of the existing legal framework on ABS in Brazil.

Whether or not there are ABS legal or regulatory requirements in place, Ethical BioTrade requires access to biological resources and associated traditional knowledge to take place with prior informed consent and be based on mutually agreed terms, with benefits being shared in a fair and equitable manner. In addition to considering more general requirements on negotiations and information exchange, these requirements mean identifying and engaging with government agencies, indigenous and local communities or other groups with recognised rights over the relevant biological resources and associated traditional knowledge. Issues to discuss include the nature of the biological resources and associated traditional knowledge, their intended and permitted uses, and safeguards in case of third-party involvement. In all cases, if traditional knowledge is used in research and commercial activities, such uses should respect the rights of traditional knowledge holders, take into account their ethical and cultural concerns, and allow their continued customary use of the traditional knowledge. For example, UEBT member TreeCrops, a small company based in Malawi, has developed a mechanism to pay a premium on *Strophantus kombe* seeds.[1] Local communities traditionally used these seeds, nowadays an ingredient of a natural heart medicine, as the basis of an arrow poison. Though research on *Strophantus kombe* took place many decades ago, the premium seeks to recognise and compensate the traditional knowledge linked to its seeds.

5.6 CONCLUSIONS

Biodiversity is a fundamental pillar for sustainable development. Businesses, especially those companies in the cosmetics sector, have an important role to play in

[1]Chris Dohse, "Use it or Lose it," CBD Business Newsletter, Vol. 5 Issue 2, May 2010, available at http://www.cbd.int/doc/newsletters/news-biz-2010-05-en.pdf.

its conservation and sustainable use.[2] Pioneering companies working with natural ingredients in the cosmetics industry, as well as those in the food and pharmaceutical sectors, are already committed to the ethical sourcing of biodiversity. For these companies, biodiversity makes business sense. It is a tool for innovation: through biodiversity, companies are able to identify and unlock more potential in their work, and add real value along the supply chain. Biodiversity also helps manage risks and secure supply chains.

Clients and consumers are increasingly demanding more information and independently-verified ethical sourcing practices from companies in the cosmetics industry. No less importantly, ethical sourcing of biodiversity can make a significant contribution to better resource efficiency and social inclusiveness. Through business engagement and practical tools and approaches, ethical sourcing practices provide guidance and support for the sustainable use of biodiversity, the respect of local communities and their traditional knowledge and practices, and the fair and equitable sharing of benefits all along the supply chain.

REFERENCES

[1] COLIPA. *COLIPA Activity Report. 2008.* Available from: https://www.cosmeticseurope.eu/downloads/136.html.

[2] Union for Ethical BioTrade. About the Union [Internet] 2011. [Accessed 2011 December 17; cited 2011 December 17] Available from: http://www.ethicalbiotrade.org/about/index.html.

[3] Natural Resources Stewardship Circle. About Us [Internet] 2011. [Accessed 2011 December 17; cited 2011 December 17] Available from: http://www.nrsc.fr/about-us/.

[4] Cosmetic Valley (2009) Our Charter for an Eco-Responsible Cosmetic Valley. Paris, France.

[5] Pimm, S.L., Russell, G.J., Gittleman, J.L., and Brooks, T.M. (1995) The future of biodiversity. *Science*, **269**(5222), 347–350.

[6] Millenium Ecosystems Assessment (2005) *Ecosystems and Human Well-being: Current State and Trends Volume I*, Island Press, Washington D.C., USA.

[7] Union for Ethical BioTrade (2009) *The Beauty of Sourcing with Respect.* Press Release 13 May 2009. Available from: http://www.ethicalbiotrade.org/dl/press/UEBT_Beauty_of_Sourcing-Web_19May.pdf.

[8] Millenium Ecosystems Assessment (2005) *Ecosystems and Human Well-being: Scenarios Volume II*, Island Press, Washington D.C., USA.

[9] Convention on Biological Diversity. Decision VI/26 on Strategic Plan for the Convention on Biological Diversity. Decisions adopted by the Conference of the Parties to the Convention on Biological Diversity at its Sixth Meeting, The Hague, The Netherlands (2002).

[10] Convention on Biological Diversity. Decision VIII/17 on Private-Sector Engagement. Decisions adopted by the Conference of the Parties to the Convention on Biological Diversity at its Eighth Meeting, Curitiba, Brazil (2006).

[2]The text of the conclusion is adapted from the script of the UEBT video on 'Biodiversity and the Green Economy', available at www.ethicalbiotrade.org/resources.

[11] Convention on Biological Diversity, Decision X/2 on the Strategic Plan for Biodiversity 2011-2020 and the Aichi Targets. Decisions adopted by the Conference of the Parties to the Convention on Biological Diversity at its Tenth Meeting, Nagoya, Japan (2010).

[12] Correa, C. (2011). *Implications for BioTrade of the Nagoya Protocol on Access to Genetic Resources and the Fair and Equitable Sharing of Benefits Arising from their Utilization*, UNCTAD, Geneva and New York.

[13] Union for Ethical Biotrade (2010) *Nagoya Protocol on Access and Benefit Sharing – Technical Brief*, UEBT, Geneva, Switzerland.

[14] Bishop, J. (ed) (2012) *The Economics of Ecosystems and Biodiversity in Business and Enterprise (TEEB) (2012)*, Earthscan, London and New York.

[15] Business and Biodiversity Offsets Programme (BBOP) (2009) Biodiversity Offsets and Stakeholder Participation: A BBOP Resource Paper. BBOP, Washington, D.C.

[16] Union for Ethical BioTrade, Ethical BioTrade Standard, STD01 – D1 – Ethical BioTrade Standard – 2011-05-04, available at www.ethicalbiotrade.org.

6

Sustainable Packaging

PART 1: INTRODUCTION

Amarjit Sahota

Packaging is becoming a focus area for many cosmetic companies looking to reduce the environmental footprint of their products. The high environmental impact of packaging is a reason for this growing focus. As stated in the introductory chapter, cosmetic packaging has been linked to death of marine life. Plastic packaging also does not biodegrade in landfills, creating environmental pollution.

It is important to note that cosmetic products have primary, secondary and tertiary packaging. Primary packaging is that which is in direct contact with the product. Secondary packaging is a container that holds the primary packaging. In the case of premium skin care products, perfumes and colour cosmetics, it is usually cardboard or other casing that houses the product on store shelves. Tertiary packaging is used for shipping, bulk handling and warehouse storage; it comprises cardboard boxes, as well as palettes for storage and shipping. The discussion of packaging impact invariably refers to primary packaging since this mostly has plastic bottles, tubs and similar packaging.

The growing prominence of sustainable packaging is evident in the sustainability plans of large cosmetic companies. Almost all such companies have made commitments to reduce the environmental footprint of their packaging. Unilever has pledged to reduce the weight of its packaging by a third by 2020 in its Unilever Sustainability Living Plan. In its sustainability plan, Procter & Gamble has committed to reduce its packaging by 20% between 2010 and 2020 [1].

There are various ways to reduce the packaging impact of cosmetic products. Apart from reduction, the other R's in the now infamous 3R's of packaging are

Sustainability: How the Cosmetics Industry is Greening up, First Edition. Edited by Amarjit Sahota.

reuse and recycle. Aside from environmental reasons, high raw material costs make it sound business sense for companies to reduce the amount of packaging they use. Thus, the drive to improve business efficiency has made packaging reduction 'the norm' in the cosmetics industry.

Companies are looking at various ways to recycle packaging materials. Glass, aluminium and paperboard have been recycled for decades; however, there is a growing drive to recycle plastics. A number of cosmetic companies are now using Post-Consumer Regrind (PCR) plastics. For instance, the UK natural cosmetics company Neal's Yard Remedies has switched all its hair care products and shower care products from glass to 100% PCR polyethylene terephthalate (PET) bottles. Aveda uses recycled aluminium for its lipstick cases while Physician's Formula Organic Wear cosmetic products are housed in cardboard containers that are partially made of recycled paper [2].

Reusable packaging is less common, however. Some cosmetic companies are opting for long-lasting bottles, compacts and tubes that can be refilled. Reusable primary packages are also becoming popular because of innovations from packaging companies. For instance, Alcan Packaging designed a pyramid-shaped sachet for Givenchy so that its perfume users can refill their 50 ml glass bottle rather than buying another.

Cosmetic companies are becoming more innovative in sustainable packaging, especially when it relates to waste management. Upcycling is gaining popularity as it involves redirecting packaging from waste streams so they can find other applications. The American company, Terracycle, works with companies like Johnson & Johnson whereby it collects used product packaging to make new eco-friendly household products that are sold in major US retailers.

The use of sustainable materials is also gaining ground. Rising petroleum prices are putting pressure on packaging companies to seek new materials, often from renewable sources. Biopolymers or plant-based plastics have become popular. One of the frontrunners is NatureWorks with its polylactic acid (PLA), which is made from corn and is fully biodegradable and compostable.

Bioplastics have low applications in the cosmetics industry. A major reason is high water permeability, preventing them to being in direct contact with many products. Their heat sensitivity also creates problems. Most applications in the cosmetics industry have therefore involved hybrid polymers. For instance, Procter & Gamble has been using hybrid polymers in its Pantene hair care products since 2012.

Other natural raw materials, such as bamboo and wood fibre, are finding packaging applications in the cosmetics industry. For example, Louvrette manufactures genuine wooden caps and refillable jars for skin care and colour cosmetic products. The US company, Physician's Formula, uses bamboo casing for its compact foundation and bronzers.

In general, there are two general approaches to sustainable packaging: design approach and materials. Both are complementary that is, ecodesign packaging can involve the use of materials with low environmental impact. The following contributors shed some light into the use of sustainable packaging in cosmetic products. The first contributor discusses the use of sustainable materials, focusing on the move from fossil-based raw materials to renewable raw materials. The second contributor discusses the role of packaging design to reduce environmental impact, whilst the third is a case study from a pioneering company involved in sustainable packaging. With the extensive amount of work done in sustainable packaging, this chapter sheds some light in this exciting new area and is not meant to be an exhaustive account.

REFERENCES

[1] Sustainable Cosmetics Summit North America, New York, May 16–18, 2012.
[2] Organic Monitor, (2010). Strategic Insights report on CSR & Sustainability in the Cosmetics Industry, London, UK.

PART 2: SUSTAINABLE PACKAGING FOR COSMETIC PRODUCTS – USING BIOBASED CARBON CONTENT AND DESIGNING FOR END-OF-LIFE

Ramani Narayan

6.2.1 INTRODUCTION

Packaging plays a critical role in almost every industry, every sector and every supply chain. However, the balance between amount and type of packaging vs. protection of product needs careful consideration – over-engineered vs. under-engineered product packaging. The goal of responsible package design is a careful balance between the amount of packaging used and protection of the product. Effective packaging makes a positive contribution to achieving a more sustainable economy for a global society.

Packaging is also a very large consumer of materials. The following materials are used in packaging:

- Metals – Aluminium
- Glass
- Paper and paperboard
- Wood
- Plastics/polymeric materials

- Hybrid constructs:
  - Plastics/polymeric materials + paper and paperboard
  - Plastics/polymeric materials + metals.

Plastic and its hybrid constructs offer considerable value for cosmetics packaging in terms of light weighting, ease of processing, rigidity or flexibility, functionality and, most importantly, product protection with affordability and durability. Indeed, about 112 million tons (40%) of the total worldwide plastics resin usage of 280 million tons is used for packaging applications [1].

Two major issues that arise from this extensive usage of plastics, especially as they relate to short-life disposable packaging, are its carbon footprint and end of life – what happens to the product after use when it enters the waste stream. This contribution describes how using biobased carbon content and designing for biodegradability and recycling can provide a value proposition to address the twin issues of carbon footprinting and end of life.

6.2.2 CARBON FOOTPRINT VALUE PROPOSITION

Carbon is the basic element that forms the basis for all plastics packaging, and therefore the issue of managing carbon in a sustainable and environmentally responsible manner becomes a critical priority. Increasing CO_2 emissions from our carbon-based products and fuels are related to the global-warming climate change issues of the day. Replacing the petro-fossil carbon in the packaging product with biobased carbon from plant-biomass resources offers the value proposition of a zero material carbon footprint. This is readily apparent from reviewing nature's carbon cycle. Nature cycles carbon through various environmental compartments with specific rates and time scales, as shown in Figure 6.2.1.

Carbon is present in the atmosphere as inorganic carbon in the form of CO_2. Photoautotrophs like plants, algae and some bacteria fix this inorganic carbon to organic carbon (carbohydrates) using sunlight for energy. This carbon sequestration to plant biomass takes place over one year (agricultural crops planted each season) to several tens of years (managed forests and tree plantations). Over geological time frames (>10^6 years) the plant biomass is fossilised to provide petroleum, natural gas and coal. We utilise these fossil resources to make polymers, chemicals and fuel and release the carbon back into the atmosphere as CO_2 in a short time frame of 1–10 years (see Figure 6.2.1). Clearly, the rate and time scale of carbon sequestration is not in balance with the use and release of carbon emissions back to the environment – more CO_2 release than fixation, resulting in an increased material carbon footprint and with it the attendant global warming and climate change problems. However, by using plant biomass, agricultural and forestry crops and residues to manufacture carbon-based products, the rate of CO_2 release to the environment at

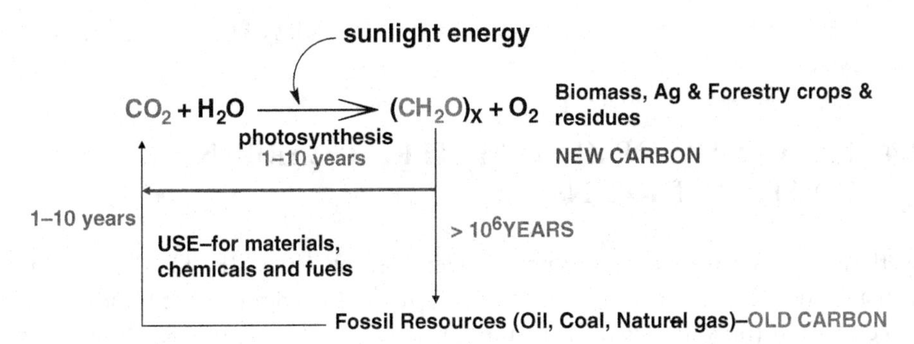

Figure 6.2.1 Illustrating zero material carbon footprint following nature's biological carbon cycle.
Source: Reprinted with permission from [2] Copyright © 2012, American Chemical Society.

end-of-life equals the rate of photosynthetic CO_2 fixation by the next generation crops or forest biomass planted – a zero 'material carbon' footprint [2, 3].

6.2.3 MATERIAL CARBON VERSUS PROCESS CARBON FOOTPRINT

So, the fundamental intrinsic value proposition of a zero material carbon footprint arises from the origin of the carbon in the product – using plant-biomass resources vs. petro-fossil resources. However, it does not address the carbon emissions and other environmental impacts for the process of converting the resource (plant biomass or petro-fossil) to product, use and ultimate disposal – the process carbon footprint. Life-cycle assessment (LCA) methodology and standards (ISO 14040 standards) [4] are the accepted tools to compute the process environmental footprint. Unfortunately, LCA focuses almost exclusively on the process (carbon and environmental) footprint. The impact of the carbon present in the product, the material carbon footprint, is treated as feedstock energy or embodied carbon energy for potential use in the next product cycle; or not at all considered, as would happen in a cradle-to-gate LCA.

It is important to calculate and report on the process carbon and environmental footprint using LCA tools and ensure that the process carbon and environmental footprint is equal or better than the process being replaced. However, the intrinsic fundamental value proposition for using biobased carbon content arise from the zero material carbon footprint in harmony with time scales of the natural biological carbon cycle.

6.2.4 EXEMPLARS OF ZERO MATERIAL CARBON FOOTPRINT RESINS

Polyethylene, polypropylene, and polyethylene terephtahlate (PET) resins are used in film and moulded bottle applications in cosmetics packaging. One hundred kg of polyolefin resin (polyethylene, PE or polypropylene, PP) contains 85.7 kg carbon. Basic stoichiometric calculations teach that the 100 kg of polyolefin (polyethylene, PE; polypropylene, PP) resin usage results in a net 314 kg of CO_2 release into the environment at its end-of-life (85.7% kg carbon present in 100 kg of polyolefin resin upon combustion will yield 314 kg of CO_2 (44/12) × 85.7). Similarly, PET (polyethylene terephthalate) contains 62.5% carbon and results in 229 kg of CO_2 released into the environment at end-of-life. However, if the carbon in PET or polyolefin comes from plant-biomass resources, the net release of CO_2 into the environment is zero, because the CO_2 released is sequestered by the next crop or biomass plantation in a relatively short time period in harmony with the biological carbon cycle time frames (Figure 6.2.2) – zero material carbon footprint. Thus, the

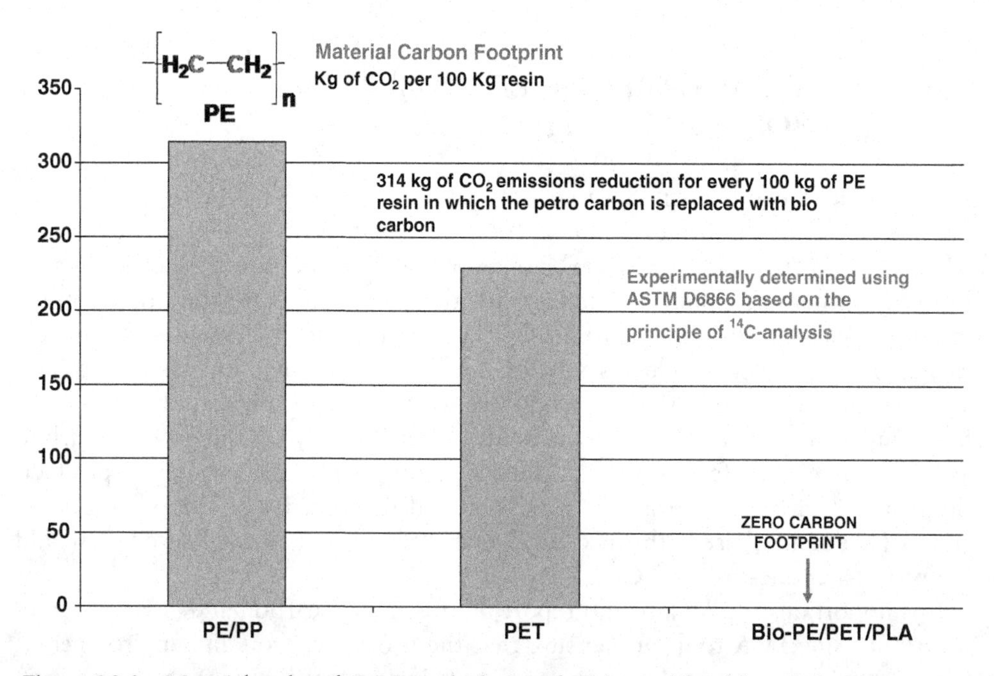

Figure 6.2.2 Material carbon footprint calculations for bio- and fossil-based plastics. *Source:* Reprinted with permission from [2] Copyright © 2012, American Chemical Society.

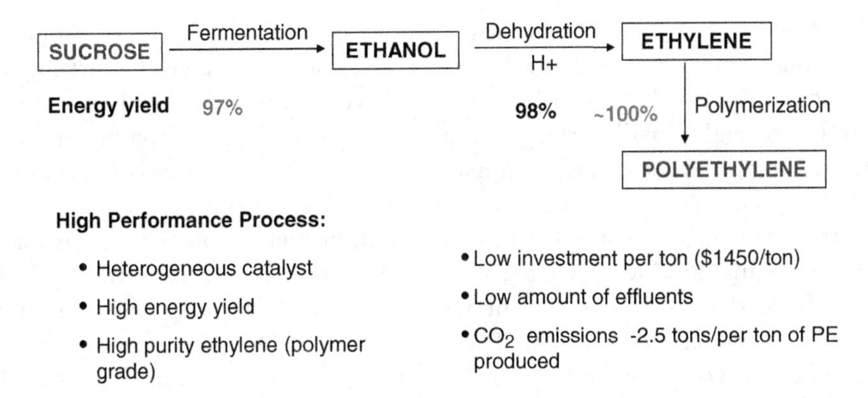

Figure 6.2.3 Braskem 100% bio-PE (zero material carbon footprint) process.
Source: Courtsey, Braskem, Brazil; ref 5.

fundamental value proposition for biobased plastics arises from this intrinsic zero material carbon footprint and not necessarily from the process carbon footprint, which may be equal or slightly better than current processes.

This approach is illustrated by Braskem, which has a 200 kton bio PE plant using sugarcane as the biobased feedstock in Brazil. Sugar from sugar cane is fermented to ethanol which is dehydrated to ethylene and polymerised to bio-PE (Figure 6.2.3). In addition, the company has a plant manufacturing 30 ktons of bio-PP [5].

Another example is the switch by Coca Cola [6] to bio-PET with 20% biobased carbon content (31.25% by mass of plant material). The PET bottle is extensively used for packaging beverages, water and a number of other food and non-food items (Figure 6.2.4). It is manufactured by condensation polymerisation of terephthalic acid and ethylene glycol. In the bio-PET the glycol component is biobased; efforts are underway to manufacture the terephthalic acid from biobased feedstocks, but currently it is made from fossil feedstock.

As shown in Figure 6.2.4, there are two biobased carbons and eight fossil carbons per PET molecule, giving total 20% biobased carbon content. On a mass basis, there

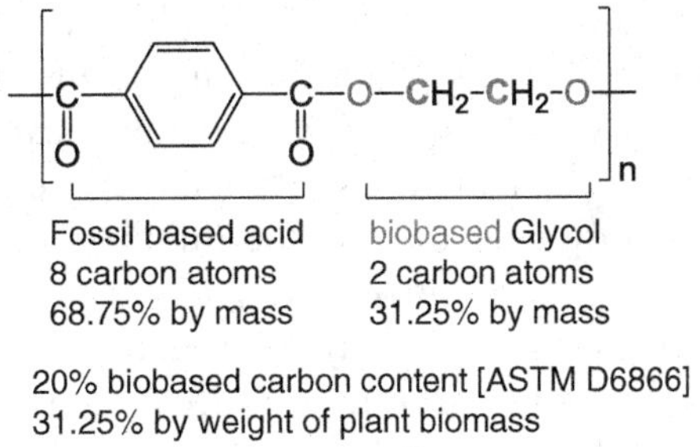

$$\left[-\underset{\|}{\overset{}{C}}(=O)-C_6H_4-C(=O)-O-CH_2-CH_2-O- \right]_n$$

Fossil based acid
8 carbon atoms
68.75% by mass

biobased Glycol
2 carbon atoms
31.25% by mass

20% biobased carbon content [ASTM D6866]
31.25% by weight of plant biomass

Figure 6.2.4 Biobased PET – Coca Cola plant bottle.

is 31.25% biobased glycol component and 68.75% terephthalic acid component. From a material carbon footprint perspective, the CO_2 emissions arising from the glycol component's two carbons would have a zero carbon emissions impact, and only the eight fossil carbons from the terephtahlic acid component would contribute to the carbon emissions impact – thus, a 20% carbon emissions reduction is achieved.

As an example, let us assume that one million metric tons of PET is used to make packaging products. As discussed earlier, two of the ten carbons in PET coming from the glycol component would have zero carbon emissions impact. Therefore, a reduction of 458 000 metric tons of CO_2 emissions is realised by just replacing two of the carbons in PET with biobased carbons. This translates to saving 1.1 million barrels of petro-oil use per year using the US EPA green house gas equivalencies calculator. Recently, major brand owners – The Coca-Cola Company, Ford Motor Company, H.J. Heinz Company, NIKE, Inc. and Procter & Gamble – announced the formation of the Plant PET Technology Collaborative (PTC), a strategic working group focused on accelerating the development and use of 100% plant-based PET materials and fibre in their products [6].

Polylactic acid (PLA) offers 100% biobased carbon content for a zero material carbon footprint, as well as both compostability and recycling as viable end-of-life options. NatureWorks LLC (a Cargill subsidiary) manufactures 140 000 tons (300 million pounds) of PLA under the trade name IngeoTM for packaging and industrial applications [7].

6.2.5 MEASURING BIOBASED CARBON CONTENT

One of the most important requirements for biobased carbon content use, and in reporting the resultant CO_2 emissions savings arising from it, is to have a transparent experimental method to unequivocally calculate the biobased carbon content in the product. As shown in Figure 6.2.5, ^{14}C signatures form the basis for identifying and quantifying biboased content. The CO_2 in the atmosphere is in equilibrium with radioactive $^{14}CO_2$. Radioactive carbon is formed in the upper atmosphere through the effect of cosmic ray neutrons on ^{14}N. It is rapidly oxidised to radioactive $^{14}CO_2$, and enters the Earth's plant and animal life through photosynthesis and the food chain. Plants and animals which utilise carbon in biological food chains take up ^{14}C during their lifetimes. They exist in equilibrium with the ^{14}C concentration of the atmosphere, that is, the numbers of C-14 atoms and non-radioactive carbon atoms stays approximately the same over time. As soon as a plant or animal dies, they cease the metabolic function of carbon uptake; there is no replenishment of radioactive carbon, only decay. Since the half life of carbon is around 5730 years, the fossil resources formed over millions of years will have no ^{14}C signature. Thus, by using this methodology, one can identify and quantify

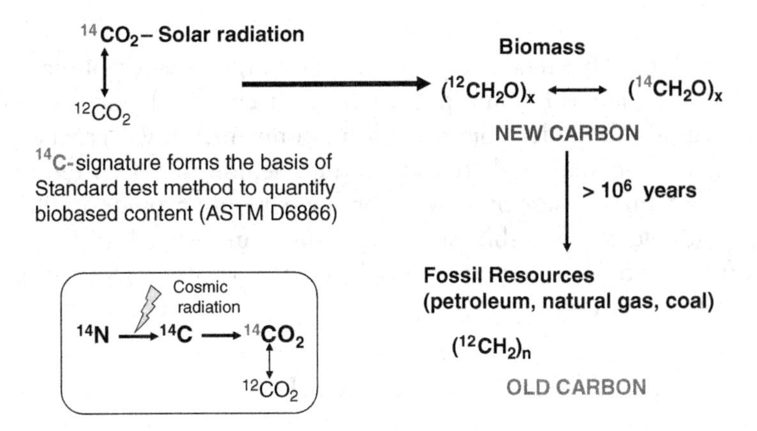

Figure 6.2.5 Measurement of biobased carbon content using radiocarbon analysis (ASTM D6866).
Source: Reprinted with permission from [2] Copyright © 2012, American Chemical Society.

biobased content. ASTM subcommittee D20.96 developed a test method (D6866) to quantify biobased content using this approach [2, 8].

The D6866 test method involves combusting the test material in the presence of oxygen to produce carbon dioxide (CO_2) gas. The gas is analysed to provide a measure of the products: $^{14}C/^{12}C$ content relative to the modern carbon-based oxalic acid radiocarbon Standard Reference Material (SRM) 4990c, (referred to as HOxII).

6.2.6 END-OF-LIFE FOR THE PACKAGING – RECYCLING AND BIODEGRADABLE-COMPOSTABILITY

Another important requirement for sustainable packaging is its end-of-life – what happens to the product packaging after use when it is disposed of.

Collectible larger bio-PET and bio-PE containers, having biobased carbon, offer zero material carbon value attribute at the 'beginning of life'. However, they are not biodegradable nor compostable. Companies using these biobased products need to have a clear strategy for collection and recycling of these products. In 2011, Europe recovered and recycled 33% of the plastics waste generated [9]. In the USA, the plastics recovered as a function of generation was 8.2% [10]. Biobased PE, PP and PET packaging offer a business opportunity for companies to institutionalise separate collection and recycling systems and provide a truly environmentally responsible brand image.

Another approach would be to design for biodegradability in targeted disposal systems, like composting and anaerobic digestion. PLA offers 100% biobased carbon content and complete biodegradability in industrial composting systems. Petro-fossil based polyesters like polybutylene adipate-co-terephthalate (PBAT), polybutylene succinate (PBS) and polybutylene succinate adipate (PBSA) are completely biodegradable under composting environment. However, the production of biobased succinic acid, and from it 1,4-butanediol, has been reported by several major companies. These biobased monomers would provide a 100% biobased and biodegradable-compostable polymer resins. Currently, blends of PBAT and PLA are offered in the marketplace as biodegradable-compostable resins with some biobased carbon content.

6.2.7 SCIENCE OF BIODEGRADABILITY [2, 11, 12]

Biodegradability is a very misused and abused term. False and misleading claims proliferate in the market. Using the term biodegradable or biodegradability is misleading unless one defines the disposal system and the time required for complete biodegradation in that disposal system. Biodegradability measures the amount of the plastic packaging's carbon consumed by microorganisms present in the disposal system as food and removed from the environmental compartment in a safe, timely and efficacious manner.

Basic biology teaches how to experimentally measure the rate and extent of this microbial utilisation; it forms the basis for all the ASTM, ISO and EN standards for measuring biodegradability. Under aerobic conditions, the carbon is biologically oxidised inside the cell to CO_2, releasing energy that is harnessed by the microorganisms for its life processes. Under anaerobic conditions, $CO_2 + CH_4$ are produced. Thus, a measure of the rate and amount of CO_2 or $CO_2 + CH_4$ evolved as a function of total carbon input to the process is a direct measure of the amount of carbon substrate being utilised by the microorganism (percent biodegradation). Claims of degradable, partially biodegradable, or eventually biodegradable are not acceptable, because it has been shown that these degraded fragments become toxin carriers further up the food chain and can have serious environmental and human health consequences. So, one has to ensure complete biodegradability in a defined disposal system, like composting or anaerobic digestion, in a short time period.

$$\text{Glucose/C-bioplastic} + 6\,O_2 \longrightarrow 6\,CO_2\uparrow + 6\,H_2O;\quad \Delta G^{0'} = -686\ \text{kcal/mol}$$

$$\text{Glucose/C-bioplastic} \longrightarrow 2\ \text{lactate};\quad \Delta G^{0'} = -47\ \text{kcal/mol}$$

$$\hookrightarrow CO_2 + CH_4\uparrow$$

It would seem obvious and logical from the above basic biology lesson that to make a claim of biodegradability, all that one needs to do is the following: expose the test plastic substrate as the sole carbon source to microorganisms present in

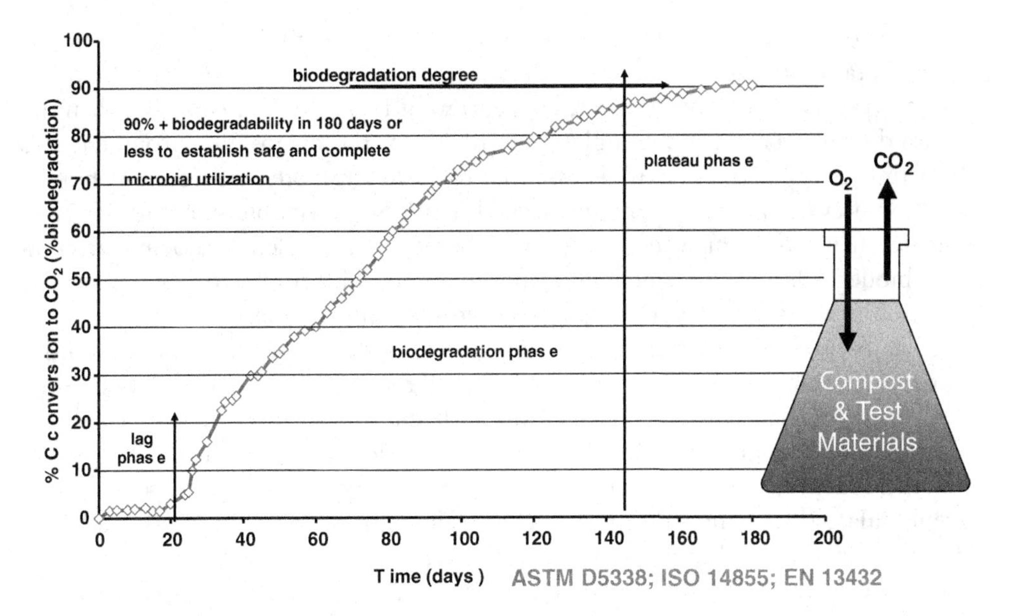

Figure 6.2.6 Experimental methodology for biodegradability using international standards. *Source:* Reprinted with permission from [2] Copyright © 2012, American Chemical Society.

the target disposal environment (like composting, or soil or anaerobic digestion or marine), and measure the CO_2 (aerobic) or $CO_2 + CH_4$ (anaerobic) evolved. A measure of the evolved gas provides a direct measure of the plastic's substrate carbon being utilised by the microorganisms present in the target disposal environment (% biodegradation). ASTM, EN and ISO test methods teach how to measure the percent biodegradability in different disposal environments based on the fundamental biochemistry described above.

Thus, one can measure the rate and extent of biodegradation or microbial utilisation of the test plastic material by using it as the sole carbon source in a test system containing a microbially rich matrix, like compost, in the presence of air and under optimal temperature conditions (preferably at 58 °C – representing the thermophilic phase). Figure 6.2.6 shows a typical graphical output that would be obtained if one were to plot the percent carbon converted to CO_2 as a function of time in days. First, a lag phase during which the microbial population adapts to the available test C-substrate. Then, the biodegradation phase during which the adapted microbial population begins to utilise the carbon substrate for its cellular life processes, as measured by the conversion of the carbon in the test material to CO_2. Finally, the output reaches a plateau when all of the substrate is completely utilised. Linear or any other form of data extrapolation from these complex biological systems is not acceptable and is very misleading because credible scientific substantiation for the extrapolation model does not exist.

Claims of degradable, partially biodegradable, or eventually biodegradable are not acceptable, because it has been shown that these degraded fragments absorb toxins present in the environment, concentrating them and transporting them up the food chain. Therefore, complete removal from the disposal environment in a short time period of 1–2 years is essential to eliminate potentially serious human health and environmental consequences [13, 14]. So, verifiable scientifically valid evidence from an approved third-party laboratory is needed to document complete biodegradability in a defined disposal system, like composting or anaerobic digestion in a short time period using the specified International standards.

It has been shown that current biodegradability data for polyolefin + additive (oxo or organic additive) formulations shows the percent biodegradation levelling off at low levels of biodegradation and not progressing further. These are additive-based plastics in which oxo and organic additives added at 1–2% levels to conventional polyethylenes (PE), polypropylene (PP), polystyrene (PS), polyethylene terephthalate (PET) and other plastics are claimed to make them 'biodegradable'. However, biodegradability claims are not substantiated by competent and reliable scientific evidence using international standards that the microorganisms present in the disposal environment are utilising the plastic carbon substrate to any significant extent (90% +) in a short and measurable time period of 1–2 years [11,12]. Unfortunately, misleading and false biodegradability claims proliferate in the marketplace and several companies have been the subject of regulatory actions.

6.2.8 SUMMARY

Using plant-biomass resources in part or completely for cosmetics packaging provides a zero material carbon footprint for the biobased carbons in the product. The biobased carbon content of the packaging can be experimentally determined using radiocarbon analysis as codified in ASTM Standard D6866. Biobased products are not necessarily biodegradable and recycling needs to be implemented for an environmentally responsible end-of-life strategy. Biodegradable-compostable plastics are not necessarily biobased and can be derived from petro-fossil resources. Major industry efforts are underway to manufacture the monomers and polymers constituting these biodegradable-compostable plastics from plant-biomass resources.

The term 'biodegradability' is much misused in the marketplace. Claims of biodegradability needs to be substantiated by reliable and competent evidence of the rate and extent of biodegradation in the target disposal environment – a graphical plot of percent biodegradability as measured by the evolved CO_2 (aerobic) or CO_2 + CH_4 (anaerobic) vs. time in days. Most importantly, if a biodegradable plastic is not completely and rapidly removed from the target disposal environment, then the degraded fragments become toxin carriers resulting in serious environmental and health risks. ASTM, European and ISO standards define and specify the requirements for complete biodegradability in composting, soil and marine environments and must be strictly adhered to so that serious environmental and health consequences can be avoided.

Biodegradability claims of conventional plastics with oxo and organic additives added at 1–2% have not been substantiated to have a significant percent (90% +) of the plastic carbon being utilised by microorganisms using ASTM or ISO standard test methods in a reasonable short time period of 1–2 years.

REFERENCES

[1] Association of Plastics Manufacturers in Europe, Brussels, Belgium; www.plasticseurope.org/.
[2] Narayan, R. (2012) *Biobased & Biodegradable Polymer Materials: Rationale, Drivers, and Technology Exemplars*; ACS (an American Chemical Society publication) Symposium Ser. 1114, Chapter 2, pp. 13–31.
[3] Narayan, R. (2011) Carbon footprint of bioplastics using biocarbon content analysis and life cycle assessment, *MRS (Materials Research Society) Bulletin*, **36**(09), 716–721.
[4] ISO 14040 (Principles & Framework) and 14044 (Requirements & Guidelines) standards, International Standards Organization; www.iso.ch.
[5] Braskem. About Braskem http://www.braskem.com.br/plasticoverde/eng/default.html.
[6] Coca Cola. About Coca Cola company http://www.thecoca-cola company.com/dynamic/presscenter/2012/06/pet-collaborative.html June 5, 2012.
[7] NatureWorks. About NatureWorks LLC; www.natureworksllc.com.
[8] ASTM International (2010) *Annual Book of Standards, Standards D 6866; D6400, D6868, D7021*. ASTM International, Philadelphia, PA, 2010, Vol. 8.03; www.astm.org.
[9] European Plastics recyclers organization http://www.epro-plasticsrecycling.org/c_124_1.html.
[10] U.S. Environmental Protection Agency (EPA). www.epa.gov/wastes; Dec 2011.
[11] Narayan, R. (2009) Bioplastics Magazine (01/09) vol. 4; www.bioplasticsmagazine.com.
[12] Narayan, Ramani (2010) Bioplastics Magazine (01/10), vol. 5 http://www.teamburg.de/bioplastics/download/Pages_38-41__from_bioplasticsMAGAZINE_1001.pdf.
[13] Thompson, R.C., Olsen, Y., Mitchell, R.P., *et al.* (2004) Lost at sea: Where is all the plastic? *Science*, **304**, 838.
[14] Thompson, R.C., Moore, C.J., vom Saal, F.S., and Swan, S.H. (2009) Plastics, the environment, and human health. *Philosophical Transactions of the Royal Society B*, **364**, 1973–1976.

PART 3: THE ROLE OF DESIGN FOR SUSTAINABLE PACKAGING

Anne van Haeften

6.3.1 INTRODUCTION

More and more brand managers and innovation managers feel the need to be sustainable. Consumers demand a more sustainable approach from their favourite brands, employees want to be proud of the products they are contributing to, society asks organisations to take a responsible role, whilst politicians are introducing laws to reward good sustainable behaviour (or discourage bad behaviour).

Working on sustainability, or working towards sustainable packaging, is a very complex and long process. The fact that there are so many roads can have a paralysing effect; the process implies cross-disciplinary thinking and a holistic view on when, how and where to start and with whom to work. People seem to forget that the 'why' is even more important than all these questions, and is at the start of the trip.

Packaging is an important part of getting to the destination on the road to becoming sustainable. Besides thinking about the structural package (material, 3D form, shape, tactile surface, etc.), graphic design can play an important role in building relationships with customers. It gives consumers a reason to believe. It's up to each brand if 'sustainable, organic, bio, eco' is something to be stated explicitly, implicitly or maybe even ... not at all.

Starting the journey towards sustainable packaging begins with a discussion with your design agency to see if they are capable of innovative design thinking! The agency will need to have structural and graphical design skills; but more importantly, they should have designers who are not afraid to think radically [1]. Besides reinventing the function of the package completely, questions will be asked and advice given about (reducing) material, waste and energy use, and at the same time about the way this should be communicated to consumers.

This section gives some approaches for sustainable packaging design. Please keep in mind that there is no ultimate solution: it depends on goals, ambitions, opportunities, markets, and so on. It is therefore impossible to present the Golden Egg!

6.3.2 THE DESIGN AGENCY

Starting the journey towards sustainable packaging means discussions with people of different disciplines at the packaging design agency (which can comprise Brand Strategists, Innovation Managers, Product Designers, Engineers, Graphical Designers and Sustainable Business Developers) but also discussions with people in your supply chain (waste management, logistics, suppliers, etc.). This contribution focuses on the role of the design agency, although working with other members of the supply chain is as (or maybe even more) important.

6.3.2.1 Being Less Bad or 100% Good

A design agency can take small steps, like minor packaging adjustments, to big steps, like a total rethink of the packaging function. Being 100% good is the basic concept of the Cradle to Cradle (C2C) philosophy of Michael Braungart and William McDonough; it creates a new way of thinking about design because it is based on the principle 'waste equals food'. It implies waste is not 'bad' but the source of new materials, new packaging or other new products. The use of alternative materials, less materials, and/or smart solutions for reuse and recycling are encouraged to develop (more) sustainable packaging.

6.3.3 PACKAGING DESIGN

> **It is all about design**
> If you start out good, you will end good
>
> Gé Moonen (CEO Moonen Packaging).

Design is the starting point when it comes to important choices for 'being good' (or being as 'less bad as possible'). If approached in the right way, design will satisfy your consumer, create a technically outstanding product and will make it economically viable. By investing in design, you can give your product personality, add emotion to it, and above all add an extra dimension to the consumer experience (and pleasure). The packaging design helps to communicate with the product's target group. A lot of innovative products have not been successful because no one thought of the essential element: the product has to look outstanding, asking to be picked up ... turned around and put in the shopping basket.

> **Adding value to your brand while reducing footprint: Good for the earth and good for business...**
> We have lived by the assumption that what was good for us would be good for the world. We have been wrong. We must change our lives, so that it will be possible to live by the contrary assumption, that what is good for the world, will be good for us
>
> Wendell Berry.

The time when 'reducing the footprint and CO_2' were the leading issues in sustainability is history. We know that creating added value in a brand is what gives consumers the reason to buy it. (Brand value: in order to create brand value, a business must be able to communicate successfully to the target market. It creates awareness of the product, and influences the consumer's perception of it. Brand value is all about the consumer recognising their need for the product, because of its perceived worth.) So we rather talk about: adding value to your brand ***while*** reducing the footprint [2] (see Figure 6.3.1).

> **Future thinking: Integrated Innovation**
> We believe the best way to effect positive environmental change is to provide consumers with a better product through design, and create mass appeal, rather than merely creating a 'green' brand that speaks only to the environmentalists. Method is hip, not hippie
>
> (Lowry, Method Products)

Figure 6.3.1 Adding value while reducing the footprint.
Source: © Reggs.

Sustainable product and packaging design are not a given. Many times, sustainability is applied to packaging design: a 'sustainable' material is chosen or the material is reduced as much as possible so the product can be labelled sustainable. Unfortunately, this approach omits many fundamental aspects of sustainability, such as function, cost reduction, producibility, assembly and consumer expectations.

The future for true innovation lies in evolution of integral design. In this evolution, all disciplines of design, be it structural, graphic or branding, are interwoven with each other from the very first stages [1]. Sustainability is an integral part in all disciplines and in all aspects of the design. Sustainability can no longer be an afterthought. For true integrated innovation, sustainability has to be self-evident.

6.3.4 THE BRAND

The design agency should understand all areas contributing to the brand (Figure 6.3.2), starting with: does sustainability fit naturally with one of your brand values, with the category? Will it answer the implicit or explicit expectations of your consumers? Can you describe their insight (with regards to sustainable packaging)? Why are they interested in the product? Are they interested in sustainability and, more importantly, might it be the main reason to buy? Are they sensitive to

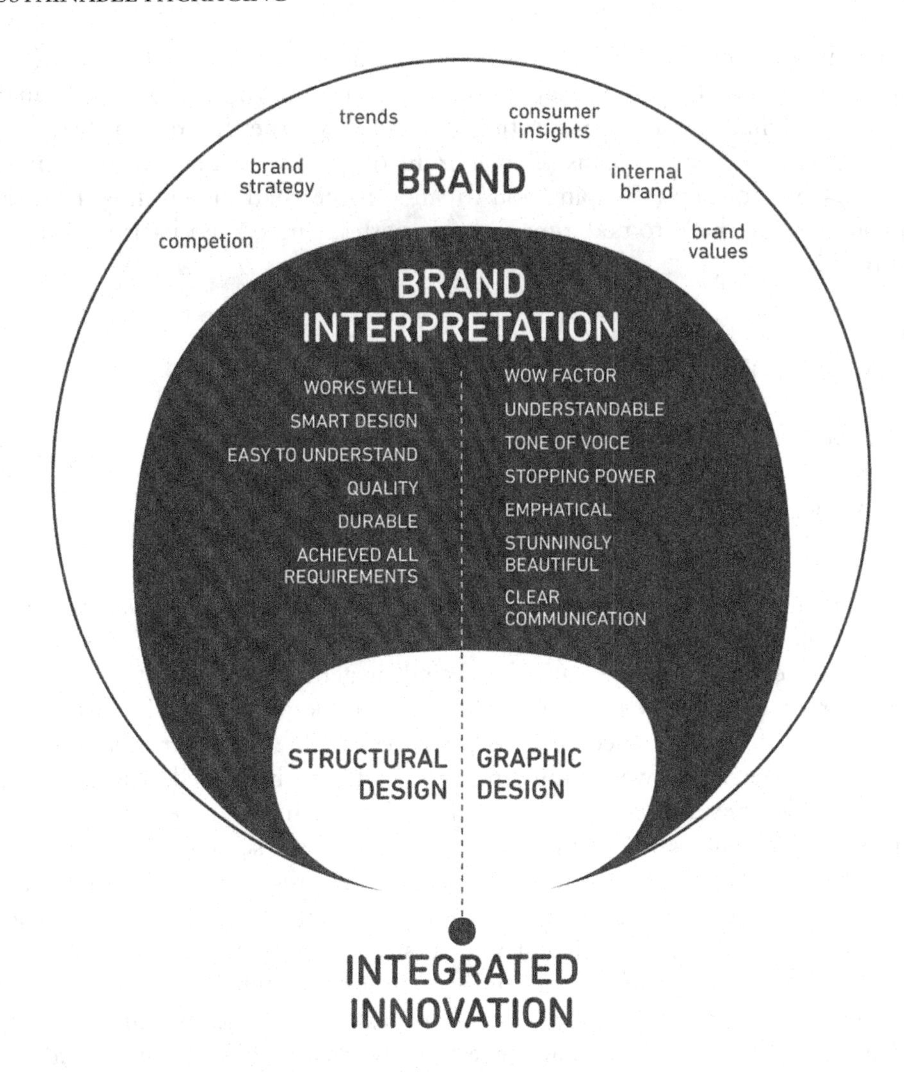

Figure 6.3.2 Integrated innovation.
Source: © Reggs.

eco-labels or . . . the opposite? Are they just looking for their A-brand because of the way the perfume smells or because of the peer group they want to belong to (image builder)?

Another thing to be taken into account is; how is your competition dealing with sustainability (communication on front of pack or not at all)? Do you want to follow this strategy or create a new vision and stand out or even have the ambition to become a new leader? Do you also want to attract a new group of consumers or only try to retain your loyal group of consumers?

The internal brand should also be taken into account: what will company staff think of sustainability? Will they be (more) proud of working on a sustainable product (and thus be ready to invest time/money) or will they feel resistance (because it will cost money and means 'change')? In other words: do you have enough support from colleagues helping you to smooth the road or will they resist and do you therefore need to take time to begin with laying a solid basis in your own company? [1]

6.3.5 INNOVATION AND DESIGN

> The inventiveness of the designer lies in a natural or cultivated and artful ability to return to those placements and apply them to a new situation, discovering aspects of the situation that affect the final design
>
> Richard Buchanan (Wicked Problems in Design Thinking).

'The best packaging is no packaging at all', is often heard in business. This is partly right; however, the basic functions of packaging should not be overlooked: product protection and preservation, especially in getting the product safely into the consumer's home. The post-consumer life of the packaging is also important [1].

Packaging designers failed to consider a second life for their creations for a good reason: packages were simply never intended to be reused. The present era of disposable convenience, which has lasted less than half a century, is quickly approaching its end – by necessity. Companies are now searching for sustainable alternatives to their printed matter and packaging; designers have an opportunity rethink the end of packaging life as nothing less than the beginning of another one (closing the loop). The biggest step to sustainability and being 100% good, cradle-to-cradle, is therefore an interesting and inspiring philosophy for designers. The cradle-to-cradle philosophy states that waste is not bad, it is useful as food for the next life. It is an idea that suggests an everlasting birth of functionality for all things designed. Designers should strive for solutions where all things, products and packaging, feed our future needs rather than compromise them. In this way, materials can be reused endlessly and recycling becomes up-cycling. These closed loops will also ensure that chemicals do not escape their productive loops to become destructive, harming mother Earth and humankind [1].

6.3.6 GRAPHICAL COMPONENT

In the beauty industry, packaging is typically beautiful, appealing and alluring. Image and status are often important (identifying with the consumer groups), so packaging design is considered crucial to the success of many beauty brands.

Structural packaging can be stunningly beautiful, inviting the consumer to touch, turn, feel and hold the product, but the graphical component (the surface) with its communication power is also a very big key to success.

How can graphic designers contribute to sustainable packaging? Graphic designers have been known to shy away from the structural challenges of package design, leaving such work to the engineers. However, designers should use their knowledge and abilities to help solve the sustainable packaging dilemma. Work together on the product personality, giving it shape, form and emotion and thus creating synergy as the two disciplines come together.

6.3.7 POST-USE PACKAGING

The most environmentally sensitive packaging can be designed, but if there's no system to collect and recycle it at the end of its lifetime, the packaging is not sustainable. End-of-life issues need to be addressed for any packaging to be considered truly sustainable (100% good). Sustainable packaging is all about systems thinking, about thinking of the whole value chain: it is not just the packaging but also about the material systems and suppliers that relate to the package. It is about waste management, as well as energy and water suppliers, and finally consumer behaviour [1].

6.3.8 LUSH CASE STUDY: GET NAKED!

Lush Fresh Handmade Cosmetics (Lush), is a UK-based company with more than 500 stores worldwide; it has been on the sustainability road from the beginning. While other brands were convinced packaging was an important marketing tool, Lush sold 65% of its products naked (without any packaging). Figures 6.3.3 and 6.3.4 show some hair care products without any packaging.

Lush uses reusable tins and paper bags and rewards customers for bringing back their containers and shopping bags. The company started using popcorn as loose filler instead of shredded paper in 2007. Lush proved it could be sustainable without higher expenses, and – more importantly – be a fresh modern brand. It used its proposition to be differentiated in the marketplace. Indeed, its 'get naked' campaign has given a lot of attention to the brand. The campaign involved its store staff serving customers wearing nothing but just aprons in July 2007. The key message of the 'get naked' campaign is that packaging is waste and thus not required.

Lush has an advantage, like the Body Shop, that not many brands have: it can use its own stores to be sustainable and deliver consumer experience at the same time. Not using any packaging and displaying its products like you would in a

Figure 6.3.3 Lush Solid Shampoo Bars.
Source: Reprinted with permission from © Lush.

delicatessen (with nice perfumes circling around the area) makes it loved by its customers.

Lush places high emphasis on sustainable packaging. Figure 6.3.5 shows some of its lipsticks packaged in clear glass bottles. Even the mascaras of Lush come in glass containers. According to the Lush web site, the company philosophy is: 'we can't

Figure 6.3.4 Lush Jungle Solid Conditioner.
Source: Reprinted with permission from © Lush.

Figure 6.3.5 Lush Liquid Lipsticks.
Source: Reprinted with permission from © Lush.

eliminate packaging completely, we look for the simple packaging to do the job and use post-consumer recycled, recyclable and biodegradable materials'. The total recycled content in its packaging is about 89%, including paper bags, aluminium tins, gift wrap, ribbons, boxes, tags and inserts. All of its pots and bottles are made with 100% PCR plastic, saving about 65 tons of C02 and 90 tons of virgin plastic, or 800 barrels of oil, each year.

6.3.9 CONCLUSION

Sustainability can create added value to brands (implying economic viability). Making small adjustments to existing products and packaging is one option to be less bad. Another option is to reach the point of being 100% good.

Think of the year 2020. Where does your company want to be, what did you achieve, what does the world look like and what role are you playing in it? Waste is no longer an issue: it is food for the next product. Energy is all green (created by Mother Nature). Toxic chemicals are no longer part of cosmetic products.

It is possible, but it needs a full rethink. After generating ideas, preferably with a diverse group (designers, brand managers, marketers, technology, corporate social responsibility managers, production, logistics, strategists, etc.), the ideas that are achievable will be distilled. Projects and project groups will be formulated and next steps planned. It is a method to take your first steps on the road, and progress step-by-step towards your idea of 100% good. Because of its complexity you will probably need a lot of time, coffee, flipcharts and energy. But it will create so much positive energy, enthusiastic colleagues, new ways of working together and above all . . . a great new future!

It's time to get started!

REFERENCES

[1] Boylston, S. *Designing Sustainable Packaging.* Laurence King Publishing, 192 p.
[2] Roscam Abbing, E. (2010) *Brand-Driven Innovation.* Switzerland: AVA Publishing SA, 192 p.

PART 4: SUSTAINABLE PACKAGING – AVEDA CASE STUDY

John A. Delfausse

6.4.1 A COMMITMENT TO THE ENVIRONMENT – THE AVEDA MISSION

Aveda's founder set the direction for the company many years ago by establishing a mission statement that guided the company both through its early years and after its acquisition by The Estee Lauder Companies years later. This mission statement is one that is repeated at corporate meetings and assemblies and is truly taken to memory and to heart by the employees of the company.

> Our Mission at Aveda is to care for the world we live in, from the products we make to the ways in which we give back to society. At Aveda, we strive to set an example for environmental leadership and responsibility, not just in the world of beauty, but around the world

Key to package design and development is this commitment: 'to set an example for environmental leadership and responsibility, not just in the world of beauty, but around the world'.

6.4.2 DIRECTION FROM THE TOP

When this contributor first started working for Aveda as Vice President of Package Development, the company president – Dominique Conseil – was approached and asked about the priorities expected in the work we were going to do. He referred to the brand equity top of mind attributes, and made performance and environmental responsibility the top priority followed by cost and design.

When the head of the organisation sets the tone for the way you work, it makes it much easier to achieve new creative and innovative solutions for your work. For the first time in my career I had not only my packaging team working towards our sustainable packaging objectives, but also the Marketing, Design, and Purchasing teams.

6.4.3 A GREAT BEGINNING

Aveda had accomplished a lot years before the packaging directive had been given by Conseil. They had developed and were using a guideline for environmental packaging which included a hierarchy of plastic materials based on their environmental impact. They were using paperboard made from 35% post-consumer pulp, and plastic bottles made with 45% post-consumer high density polyethylene. They were working with their tube supplier to develop the first extrude plastic tube with post-consumer resin. The natural next step was to further develop these packaging forms to increase the levels of recycled material.

6.4.3.1 Rosemary Mint Shampoo

Rosemary Mint Shampoo had been packaged for years in a classic Boston round bottle with the Aveda logo embossed on the bottom of the bottle. Over the years, Aveda had taken the plastic resin recycled content from 25% to 45%, and finally to 50%. This was unheard of in the cosmetics and hair care industry. California legislation had required plastic bottles to have a minimum of 25% Post-Consumer Recycled (PCR) content for years, but beauty care products always had an exemption. When the brand decided to redesign the bottle in 2001, they were looking to make the shape more contemporary and attractive to their consumer. To accomplish this, they designed a slimmer and taller bottle with the Aveda logo now embossed on a band around the neck of the bottle. Of course, they kept the Boston round shape which guaranteed continuity of brand recognition (see Figure 6.4.1). What made the introduction of this bottle even more incredible was the ability to increase the PCR level from 45 to 80%, and at a lower cost. The PCR level was able to be increased by adding PCR to the outer level, as well as the inner level of the co-extruded bottle. Although the increase in PCR added a grey tone to the resin, it was basically indistinguishable to the consumer because product contact with the surface of the bottle hid the colour impact. Most of the products in the hair care line of Aveda are also in coloured bottles, which totally hides the impact of the PCR.

When Dominique Conseil was approached about this difference, he agreed that it was the right thing to do. All of Aveda's high-density polyethylene hair care bottles now contain a minimum of 80% PCR. But talk about taking a leadership role: Aveda has now started to introduce bottles with 100% PCR. In order to do so, Aveda had to develop the capability to carry the bottle colour in PCR resin: another

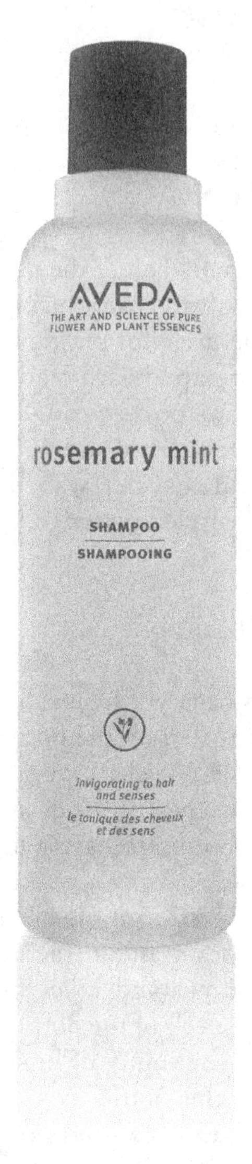

Figure 6.4.1 Aveda Rosemary Mint Shampoo.
Source: Reprinted with permission from © Aveda.

first for the industry. It cannot be stressed enough that the ability for a brand like Aveda to make this kind of innovative advance is due to the commitment to its mission from the top.

6.4.3.2 Why PCR?

Although using PCR is a challenge due to the quality, colour and consistency of material collected and reprocessed, Aveda has always specified post-consumer recycled materials when talking about recycled content. This sets their packaging goals to a higher standard, but also has the effect of sometimes making their claims about their packaging appear less impactful. For example: another brand could be using 25% PCR and 75% pre-consumer material and making a claim of 100% recycled, while Aveda makes a claim of 80% recycled PCR. Why PCR? The request for PCR builds a market for the material and encourages the development of the technology and infrastructure that is needed to actually recover materials once they have been used for their intended purpose. Using the material also requires the development of innovative material reprocessing methods as well, to insure a steady stream of clean and usable feedstock. The brand also feels that industry should do its part to optimise material use in their processes.

6.4.3.3 Aveda Brilliant

In 2002, Aveda decided to re-launch its Brilliant styling line. This gave the opportunity to not only improve the environmental impact of the packaging, but also to introduce the first PCR tube. The Aveda Package Development team was at that time using their plastic hierarchy list to recommend materials for bottles high-density polypropylene (HDPE) was at the top of the list and, therefore, the recommended material for the re-design. The old Brilliant line, however, was packaged in blue glass that was translucent. That level of transparency was never going to be achieved with HDPE. Somewhere down the list was another material, PET. PET, although not as environmentally desirable, offered a look very similar to glass. So the challenge was to maximise the PCR level of the PET to achieve the environmental edge required, as well as satisfying the marketing and design requirements.

Several levels of PCR were tested before a bottle was successfully made from 100% PCR PET. For the first time, a brand was able to get to 100% PCR. The bottles were also considerably lighter than the glass and did not require a secondary folding carton. The gold leaf decoration was eliminated in favour of white silk screening and a wrap label which carried all the required legal copy.

In the end, a so called win–win–win solution was obtained. Although the priorities were performance and environmental responsibility first, followed by cost and design, all four were achieved. Packaging was updated to meet consumer needs, whilst making a considerable environmental improvement; at the same time, significant reductions were made in the annual cost of goods per product sold. This was similar to the success the team had had the year before with Rosemary Mint.

6.4.3.4 Brilliant Tubes

As mentioned above, with the relaunch of the Brilliant styling line, a truly innovative tube was introduced to the industry. For the first time, the HDPE PCR was captured in the middle layer of a co-extruded tube. Aveda had been working with a supplier for many years to achieve this success. The tube was the first of its kind containing linear low-density polyethylene (LLDPE) in the inner and outer layers of the triple-extruded tube sleeve, while adding HDPE PCR to the middle.

At first, the tubes would not hold their round shape at the opening and were impossible to fill. The extruder speed had to be slowed down to get a good tube. Other issues popped up during the three years of development, but it all ended well in 2002 with Brilliant.

This innovation is a good example of how it sometimes costs more to do the right thing. By slowing down the extrusion speed, we incurred added costs by producing fewer tubes per shift. Although the PCR actually cost less than the virgin material, the overall cost ended up higher due to processing differences. In later years, Aveda's supplier was able to get back up to normal production rates by designing specific extruders to meet this new requirement for PCR in tubes.

6.4.3.5 Uruku Lipstick

One of the major redesigns for Aveda, and a major environmental breakthrough, came in 2003 when the brand decided to repackage their make-up line. The design was to use a square to round profile for the containers and maximise PCR content. At first, there were thoughts of going with the preferred material, HDPE. However, there were no lipstick cases made from this material and if one was going to be designed with new tooling, as well as doing all the testing necessary, it would have taken many years to get the product to market.

Instead, the possibility of using the basic lipstick cartridge in the industry was considered, substituting PCR materials. This was a real challenge! Although PCR PET and HDPE are readily available from beverage bottles and milk bottles, recycled polypropylene (PP) and high-impact styrene (HIS) were not to be found, especially as PCR. These were the basic materials that made up most lipstick cases. This also gave the opportunity to use the cartridge as a refill unit and match it up with a luxury outer accessory case to make the total lipstick package, much like razor blades and handles.

It cannot be stressed enough how important Aveda's supplier was in making this a success. Without their willingness to take risks and go where no one had gone in the past, it would never have been possible to accomplish all that was done in the packaging. The supplier not only adjusted all their tools to enable the refill case and the moulding with PCR materials, they also sourced PCR PP and HIP for Aveda.

For the outer carton, the team also went another mile by substituting a moulded pulp clamshell for the normal folding carton. Moulded pulp is made from 100% old newspapers. This cost a bit more in the processing, but added that extra touch of creativity and stretched the envelope one more time.

Needless to say this award-winning lipstick was one of the highlights in this contributor's tenure at Aveda.

6.4.4 REAL SUSTAINABILITY

Somewhere along the line, it was felt that, although Aveda was a leader in the use of packaging with recycled content, there had to be more to being environmentally responsible. The breakthrough came when I was asked to contribute to an issue of *Packaging World* as a follow up to an article about Bill McDonough and Michael Braungart's book *Cradle to Cradle*. This led to a meeting at the University of Virginia which led to the formation of the Sustainable Packaging Coalition and its Definition of Sustainable Packaging. For the first time there was a good idea of what needed to be taken into consideration when designing packaging.

Goals of Sustainable Packaging:

- Is beneficial, safe and healthy for individuals and communities throughout its life cycle;
- Meets market criteria for both performance and cost;
- Is sourced, manufactured, transported, and recycled using renewable energy;
- Optimises the use of renewable or recycled source materials;
- Is manufactured using clean production technologies and best practices;
- Is made from materials healthy throughout the life cycle;
- Is physically designed to optimise materials and energy;
- Is effectively recovered and utilised in biological and/or industrial closed loop cycles.

Source: Sustainable Packaging Coalition.

6.4.4.1 Renewable Energy

The use of renewable energy is just one way to demonstrate how Aveda has taken this definition to heart and stretched their understanding of sustainability. As a company, Aveda made a commitment to buy renewable electric power credits to offset the 100% of electricity used in its primary manufacturing facility, and purchases Green-e certified renewable energy offsets to balance natural gas use. Not to be outdone, the packaging team challenged their suppliers to do the same. The first supplier on board was Johnson Printing. They found that if they were to buy renewable energy, the added cost was very little after taking into account the savings of a lighting upgrade they had just gone through. For the first time, Aveda was able to not only responsibly source recycled board, but also manufacture the board, convert that board to a folding carton and then fill that carton in its plant using renewable energy.

Not all the points in the definition can or need to be achieved at once, but the journey to sustainable packaging has a lot of opportunities along the way to build on each achievement and improve a company's overall impact on the environment.

A detailed account of Aveda's renewable energy and waste management programme, as well as additional details on sustainable packaging, are given in the following Chapter 7.

7

Energy and Waste Management

Charles J. 'Chuck' Bennett and Michael S. Brown

7.1 INTRODUCTION TO ENERGY AND WASTE MANAGEMENT IN THE COSMETICS INDUSTRY

7.1.1 Global Resource Constraints and the Challenge for Business

We live in a world where the natural resources we use in our economy are finite and where the capacity of the Earth to absorb the wastes and emissions associated with producing and using those resources is limited. This premise is not universally accepted and has been the subject of much debate in recent decades. In 1980, economist Julian Simon and biologist Paul Ehrlich entered into a bet over Ehrlich's claim (echoing Thomas Malthus) in his 1968 book, *The Population Bomb*, that the world's population was expanding more quickly than growth in food and resources and that ecological disaster was just around the corner. The bet was whether the price of five commodity metals would rise or fall by 1990; Ehrlich expected growing scarcity to result in higher prices and Simon believed that human ingenuity would drive technological change resulting in lower prices. Ehrlich famously lost the bet – all five metals dropped in price (adjusted for inflation) – and commentators had a field day with the notion that humanity could safely ignore any limits on resource use [1].

But there's a different interpretation of the bet. Simon wasn't really arguing that resources would never be used up. What he was saying is that when a resource

Sustainability: How the Cosmetics Industry is Greening up, First Edition. Edited by Amarjit Sahota.

gets too expensive, humans would find alternatives (think fiber optics substituting for copper cables, or deep-sea bed drilling substituting for shallow land wells) that would result in prices for that resource coming down. And while Ehrlich took the bet, he wasn't thinking like an economist (that price signals encourage innovation), but as a biologist who saw that demand for individual resources both depletes those finite resources and creates pollution (externalities that aren't included in the price of commodities). In effect, even though Simon won the wager, both the economist and the biologist were right [2].

Our economic system does a poor job of accounting for the environmental harms associated with producing, using, and discarding natural resources, whether biological in origin or fossil fuels or inorganic. This complicates decision making for businesses. For example, the usual tools for calculating return on investment do not incorporate the global warming "costs" of cheap energy such as coal-generated electricity or the environmental benefits of using more expensive photovoltaic-produced electricity. Nor do these tools discern the full value of a return on investment in establishing closed-loop systems for maintaining control of materials through production, distribution, and consumer use cycles to minimize constant dependence on virgin materials.

It is in the context of these challenges that businesses generally – and increasingly the cosmetics industry – are developing innovative approaches to business to address complex sustainability challenges. In the remainder of this chapter we focus first on the complexity of the issues surrounding energy and waste management generally in the cosmetics industry. Then we shift attention to one company – Aveda – and focus on how this industry leader has evolved in its approaches to innovative energy and waste management practices and, via its mission-driven commitment to care for the world we live in, set an example for environmental leadership and responsibility.

7.1.2 Energy Issues and the Cosmetics Industry

Energy is a key resource for the use of a wide variety of inputs that go into making personal care products. For synthetic ingredients, fossil fuels (crude oil and natural gas) are the raw material inputs for the specialty chemical processing that serves the bulk of the industry. Extensive energy is required during the life cycle to discover, extract, process or refine, and distribute oil and natural gas and process these into synthetic ingredients. Much of the energy consumed is from the same non-renewable resources used to make the ingredients along with coal; although in some locations, a portion of electricity generation (critical for motors and pumps) comes from renewable sources (wind, solar, geothermal) as well as large hydropower and nuclear generation facilities. A variety of substances that are required in the production of oil and gas such as drilling "muds" and "fracking" fluids and the chemical substances involved in the production of synthetic ingredients all require energy resources to produce and use.

Similarly, the production of plant-based ingredients, such as essential oils, colorants, surfactants, emulsifiers, and moisturizers, incorporate energy throughout the production cycle. Farms use energy directly for farm machinery, transportation, and irrigation. Although the plants used to create these ingredients may be grown by relying on rain as the sole water source, many are produced with irrigation that diverts water by pumping from ground and surface water sources. Energy is also an indirect factor on the farm; much is embedded in the production of agricultural chemicals such as pesticides, herbicides, and fertilizers; pest control and fertilizers suitable for organic production are dependent on energy use as well.

Much of the energy consumed on farms is from non-renewable resources (fossil fuels), although in some locations, a portion of electricity generation (critical for motors and pumps) comes from renewable sources (wind, solar, geothermal) as well as large hydropower and nuclear generation facilities. Key fertilizers are often produced from fossil fuel sources (e.g., nitrogen); phosphates are an energy- and water-intensive fertilizer that is at risk of being considered non-renewable as easily accessible deposits are fully exploited [3]. Organic agriculture reduces the fossil fuel content of some of these resource needs, but does not entirely eliminate it.

Beyond the farm, energy is a critical input across the life cycle of personal care products, such as the use of distillation to obtain essential oils from various plant materials or to transform raw plant material into a surfactant. Manufacturing cosmetics requires energy to transport ingredients from suppliers around the world to production facilities via rail, truck, boat, and occasionally air. Production processes vary. In some cases, products can be manufactured using ingredients that are soluble at ambient temperatures and required energy is limited to mixing and transferring from batch to bottle. Other types of products, particularly emulsions such as creams, lotions, and conditioners, contain high melt-point materials that require heat processing and sometimes flash cooling, substantially raising the amount of required manufacturing energy.

Virtually all personal care products are delivered to the consumer in packaging. Liquids are typically delivered in a watertight container; other products may be packaged in tubes, jars, or boxes. Common packaging materials are biobased (paperboard), fossil fuel-based (polymers such as polyethylene, polyester), and inorganic (glass). All of these require energy, whether embodied in the machinery and agricultural chemicals used to manage and harvest forest resources, the raw materials substances involved in the production of polymers (synthetic, regenerated, or manufactured), or the furnaces to melt the sand, soda ash, limestone, and recycled glass that are the typical raw materials for new glass.

Beyond manufacturing, fossil fuel energy is central to transportation of cosmetic products to points of sale or professional use. For some categories of products (e.g., shampoos) energy is consumed to heat water for shampooing at a salon or in the shower at home, while other categories such as skin care typically do not require energy in use.

7.1.3 Wastes and Personal Care Products

From the farm and wellhead to the manufacturing facility, the retail outlet, the salon, and the household, wastes and emissions are a major economic and environmental cost for the personal care product industry. At the farm, wastewater run-off carries pesticides, fertilizers, and sediment to surface and groundwater potentially polluting drinking water and causing eutrophication of receiving waters (high plant mass growth resulting in depletion of oxygen) [4]. Drilling muds and fracking fluids in wastewater streams at oil or gas well sites may also contaminate local water supplies. Processing of farm products may result in wastewater that contributes to eutrophication and toxicity. Personal care product manufacturing requires strict attention to cleanliness, resulting in wastewater streams that may also require treatment before discharge to receiving waters.

Particulate air pollution from soil erosion, pesticide drift from applicators, hazardous air pollutants from oil and gas production, and general air pollution from the production of electricity and the burning fossil fuels are all part of raw material and personal care product processing. Of particular significance is the role of energy – especially fossil fuels – in climate change and global warming processes.

Solid waste is generated across the life cycle of personal care-related production processes. Energy-related solid waste is in the form of mining and drilling residues. At the farm, pesticide and fertilizer containers are the majority of solid wastes. In manufacturing, raw ingredient packaging (bags, pails, drums, totes, liners, pallets) and the packaging for consumer containers (corrugated boxes and paperboard) are typical wastes. Manufacturers ship products in corrugated boxes to distributors and retail stores; products are typically removed from the boxes for display on shelves and the shipping boxes become potential waste. Virtually all consumer packaging – bottles, tubes, applicators, jars, boxes, and so on. – may end up as solid waste following consumption of the product.

Hazardous wastes may be generated throughout the life cycle of personal care products. Residual and unused pesticides, herbicides, and related farm chemicals, the hazardous substances endemic in oil and gas production and the chemistry of synthetic polymers, hazardous solvents used in the processing of some plant-based and fossil fuel-based ingredients, and some cleaning chemicals used in manufacturing operations may require management as hazardous waste when use is no longer possible.

It is in the context of the preceding discussions of resource constraints that Aveda sees its role as an innovator willing to take chances on new technologies in an effort to align its business practices with the needs of the planet. In doing so, Aveda has sought to understand what it means to do business in a resource-constrained world and to develop its business in a manner that is consistent with the short-term realities of a highly competitive marketplace *and* long-term planetary health. Aveda has strived to deliver consistent high-quality products while pushing the envelope with novel plant-based ingredients, promoting organic agriculture in its supply chain, industry leading levels of recycled content, and other innovations in product packaging. A critical dimension of all of Aveda's business practices has also been innovation in the use of energy and the elimination of wastes.

7.2 AVEDA – THE COMPANY

Aveda has been committed to Earth and Community Care – Aveda's term for environmental sustainability and social responsibility – since the company was founded in Minneapolis, Minnesota in 1978 (Figure 7.1). This commitment has been holistic, from the development of products derived from plants that are beneficial for users – both professionals and individual consumers – through sourcing of raw materials, to manufacturing and distribution operations, and to Aveda's commitment to philanthropy. Aveda is a mission-guided company (see Box 7.1) and this mission informs company decision making and actions. The key themes of the mission are "care" for the earth and society and "environmental leadership and responsibility."

Figure 7.1 Aveda's headquarters and primary manufacturing site in Minnesota. The site is certified wildlife habitat and storm water is captured and used for lawn irrigation.
Source: Reprinted with permission from © Aveda.

Box 7.1

Our Mission at Aveda is to care for the world we live in, from the products we make to the ways in which we give back to society. At Aveda we strive to set an example for environmental leadership and responsibility, not just in the world of beauty but around the world.

Aveda's founder, Horst Rechelbacher, was committed to environmentally responsible manufacturing of these products. He was an active environmentalist throughout the 1980s and led Aveda to become the first privately held company to sign the Valdez Principles (later to become the Ceres Principles) in 1989. Two of these principles helped guide the company's early efforts on energy and waste:

- **Energy Conservation.** We will conserve energy and improve the energy efficiency of our internal operations and of the goods and services we sell. We will make every effort to use environmentally safe and sustainable energy sources.
- **Reduction and Disposal of Wastes.** We will reduce, and where possible eliminate, waste through source reduction and recycling. All waste will be handled and disposed of through safe and responsible methods.

During the company's early years, much of the focus on energy and waste was limited to using recycled packaging when possible as a means of encouraging public recycling, since Aveda was manufacturing a limited range of products and depending primarily on third-party manufacturers.

This situation changed in 1991, when Aveda moved into its current manufacturing facility in suburban Minneapolis and repurposed a former National Cash Register plant to produce personal care products. With this transition, Aveda began to focus on improving energy efficiency and reducing waste both in its operations and in its products and packaging, trends that gained momentum through the 1990s and beyond. During this period as well, Aveda broadened its portfolio of products beyond its principle focus on hair care to include a growing range of skin care lotions, creams, and cleansers. In conjunction with its expanding skin care line, Aveda faced novel manufacturing challenges different from those associated with hair care products. Also, Aveda found that increasingly sophisticated products in all categories can require more energy inputs and often more intensive cleaning processes, all of which contribute to increasing energy demand and wastewater streams.

At the end of 1997, realizing that the future success of Aveda required more comprehensive business resources and scale, Rechelbacher sold Aveda to The

Estée-Lauder Companies, Inc. (ELC). ELC has recognized and consistently supported the contribution of Aveda's strong public sustainability commitment to the value of the brand. They hired Dominique Conseil, a veteran of the personal care industry with a very strong environmental and social responsibility approach to business, to lead the brand in 1999; that leadership has continued to the present.

In 2001, Aveda developed its formal environmental sustainability policy that publicly committed the company to operate in an environmentally sustainable way, including commitments to reduce the impacts of climate change and to eliminate waste. Together, the Aveda mission, the environmental sustainability policy, and the Ceres Principles guide Aveda's sustainability path. They have led to steadily improving energy efficiency, increasing use of renewable energy, and to purchasing greenhouse gas (GHG) emissions offsets where renewable power cannot be directly substituted. They have also driven Aveda to increase rates of recycling and materials reuse in operations and to establish an industry-leading commitment to packaging innovation that is part of a broader pledge to incorporate cradle-to-cradle (C2C) principles both in the design of Aveda products and packaging and in its overall operations.

Aveda's achievements in the areas of energy and waste have been at the forefront of the cosmetics industry, but much remains to be done as the needs of the planet and society become ever more critical and the Aveda business continues to grow. The subsequent sections of this chapter will focus on both achievements and opportunities.

7.3 ENERGY MANAGEMENT IN AVEDA

"Managing" energy in the business ideally is a comprehensive process where the goal is to influence or control energy in all stages of a product's life cycle from raw materials to ultimate product uses. Realistically, this is far easier said than done and most businesses focus – especially initially – on what they can most influence and control – operations and certain components of their value chains. This has been Aveda's experience.

The basic path that Aveda has followed has been to reduce energy use and increase the efficiency of its use, then purchase renewable energy through the state electrical grid when it became available, purchase offsets for natural gas use and electricity used in operations where renewable energy is not readily accessible and, most recently, to begin to produce solar energy on site. Aveda's progression over the past 20 years is one that any business can follow and, with the knowledge and technology that has been developed during this period, this path can be significantly shortened today.

7.3.1 Process Energy Opportunities

Aveda focused on both process energy efficiency and building energy efficiency from the outset. In the process category, one early effort was to reduce the need for heating and cooling in manufacturing by pursuing development of product formulae that could be processed at ambient temperatures. While this was a successful strategy with many early formulations, use of a growing range of ingredients in an increasingly diverse product portfolio resulted in a steady increase in heat requirements for product processing.

A second strategy in manufacturing that led to reduced energy requirements is batch timing and the number of batches that need to be produced. In the early days of manufacturing, Aveda produced a relatively small number of products in large batches using relatively short processing times compared to the present state of manufacturing. As the business has grown, however, the number, complexity and total volume of batches and products have increased and this has led to increased processing times and additional change-overs with a greater need for cleaning and attention to sanitation, all of which have resulted a steady increase in energy demand in operations. Offsetting the increase is a number of efficiency improvements including:

- Upgrading to high efficiency electric motors.
- Using insulated jackets on some heated processing tanks.
- Clean-in-place systems and other sanitation efficiencies.
- Using a closed-loop chilled-water system to reduce the energy used for cooling.
- Installing high efficiency boilers for steam production.

Energy efficiency upgrades have been a continuous process over Aveda's nearly 20 years of manufacturing and have contributed – along with building energy efficiency improvements described below – to a steady decline in the amount of energy *that would have been needed* without these improvements. Harvesting this significant "low-hanging fruit" has resulted in a fairly stable level of "normalized" energy use (energy per kg of production), even with steady increases in the annual volume of production. However, these improvements have not offset continued growth in the total absolute energy needed to meet increasing production demands.

7.3.2 Facility Energy Improvements

Improving energy efficiency in Aveda's primary facility in Blaine, Minnesota has had two primary emphases; lighting and heating, ventilation and air-conditioning (HVAC). Aveda joined the US Environmental Protection Agency (EPA)'s Green

Lights program soon after moving into the facility in Blaine and began systematically upgrading lighting with higher efficiency lamps, timers and motion sensors, and other technologies that have become widely used in building energy management. This process has continued over time with multiple iterations of lighting upgrades, for example. Currently, Aveda is experimenting with light emitting diode (LED) lighting and has plans to convert to LEDs where appropriate as lighting systems need to be replaced.

In addition to these upgrades, in 2006 Aveda took advantage of the opportunity to install skylights with solar powered mirrors that follow the sun and maximize reflected sunlight into the interior of a newly constructed distribution center built next to its existing manufacturing facility, significantly reducing the need for artificial lighting during the daytime. Aveda has consistently sought out and used the extensive resources available through government agencies, energy utilities, and the US Green Building Council, to guide its energy-efficient new construction retrofitting of existing buildings.

As with lighting, Aveda embarked on a continuous improvement process to upgrade energy efficiency in building heating and cooling. An early innovation – in 1994 – was the construction of a south-facing solar wall on the roof of the manufacturing facility which results in pre-warmed air being drawn into the HVAC system to reduce the heating system load in the winter. Other innovations have included:

- Computer-managed heating and air conditioning systems.
- A closed-loop HVAC water system.
- Air-to-air heat exchangers that recoup energy losses to ventilation.
- High-efficiency SEER rated heat pumps.
- A night-time air flush system in the distribution center eliminating the need for air conditioning except in office spaces in the building.

7.3.3 Results and Current Situation

Aveda has significantly decreased energy consumption in its direct operations over what it would have been through sustained, consistent search for efficiency opportunities. Between 1994 and 2009 – the last year for which complete data are available – cumulative energy savings were:

- Electricity: Over 56 million kWh.
- Natural gas: Nearly 850 000 therms.

These reductions saved Aveda over $3.4 million in energy costs. Over the 15 years of this process, it has become increasingly difficult to identify and financially justify

Energy Type	2008	2011	Percent Change
Natural gas total use (MJ* – millions)	21.4	21.2	−0.5
Natural gas normalized (MJ/kg production)	2.37	2.18	−6.4
Electricity total use (MJ – millions)	26.8	21.3	−14
Electricity normalized (MJ/kg production)	2.97	2.37	−19.1
Total energy use	48.2	42.5	−12.0
Normalized energy use (Mg/kg production)	5.34	4.55	−15.0
Total production – kg	9,167,985	9,743,993	6.3
Total production – units	40,391,704	47,689,496	18.1

Figure 7.2 Recent improvements in Aveda energy efficiency compared to production changes.
*megajoules
Source: Reprinted with permission from © Aveda.

additional efficiency improvements. The low hanging fruit have all been picked and we are searching the larder to reach higher shelves for additional energy savings (Figure 7.2).

Aveda expects to achieve the next level of efficiency improvements by installing more granular metering systems to better understand energy use at the individual process and machine level and by using continuous monitoring of batch production processes to ensure that energy use does not exceed levels specified for different processes and process stages. Aveda believes that this level of control over energy use will be essential for achieving ongoing efficiencies and that a portion of the anticipated payback from these investments will offset by expected increases in energy costs. Aveda anticipates a significant upswing in production at its Blaine facility as a result of growth in future Aveda product sales as well as providing expanded production capacity for additional Estée-Lauder brands, further underscoring the need for increasingly sophisticated energy management to achieve its sustainability commitments.

7.3.4 Renewable Energy and Emissions Offsets

In 2006, Aveda decided to go beyond energy efficiency and invest in renewable energy to reduce the company's carbon footprint, consistent with its environmental sustainability policy. The immediate opportunity for doing so was offered by the electrical utility in Minnesota through its WindSource® program. Aveda pays a slight – but only slight – premium for electricity, which supports the development of wind turbines in the utility's system equivalent to Aveda's demand. In this way Aveda's consumption of electricity is replaced by an equivalent amount produced by wind turbines in Minnesota. All of Aveda's owned-facilities in Minnesota purchase WindSource power.

Aveda Emissions Offsets and RECs: FY 2009–FY 2011

- Participated in the incremental funding of the Greensburg, KS wind farm following in the aftermath of the EF-5 tornado that leveled the town
- 5000 short tons of VERs (verified emission reductions) produced during 2009 from bio-mass energy projects in Brazil
- VERs produced during 2009 from wind energy projects in India.
- 5000 short tons of VERs produced during 2010 from the Clinton County Landfill Gas Methane Project
- 2500 short tons of VERs produced during 2011 from the Wewoka, Oklahoma Landfill Gas Methane project
- 7.3 million kilowatt hours (kWh) of renewable energy credits during 2011

Figure 7.3 Summary of Aveda's Purchases of Emissions Offsets and Renewable Energy Credits (RECs) to Balance the Emissions of our Operations.
Source: Reprinted with permission from © Aveda.

In addition to WindSource energy in Minnesota, Aveda purchases carbon offsets and renewable energy credits (RECs) for energy used in its owned and leased facilities – including approximately 120 retail stores – in the USA and in some overseas locations, as well as for natural gas use at its Minnesota facilities. Offsets and RECs are obtained through Native Energy, which conducts due diligence and guarantees their quality. Aveda "retires" the offsets with Clean Air–Cool Planet to ensure that they are not available for resale and thus become potentially meaningless (Figure 7.3).

The long-term energy goal is to produce sufficient renewable energy on Aveda's sites to meet operational needs, eliminate dependence on market sources for energy, and potentially sell renewable energy into the commercial grid. None of these goals are likely to be realized soon. Aveda's ability to produce renewable energy is constrained by both geographic conditions (wind and solar resources) as well economic conditions (e.g., costs for solar in Minnesota including tax benefits and subsidies are still prohibitive compared to the costs of utility-provided energy available with the cost of offsets included).

Nevertheless, Aveda has embarked on a path toward this goal with the installation of an initial solar energy array at its California distribution center that produces about 50% of the energy needed for that facility and a pilot solar installation at its Blaine, MN facility that is currently being monitored to determine the potential for solar given the Minnesota climate conditions, space available for arrays, and cost. For the time being – with current technologies –production of wind energy at Aveda's owned locations is impractical due to low wind potential.

Elsewhere, most Aveda facilities are in leased shopping mall locations and are thus dependent on other parties for either potential upgrades in efficiency or to use onsite renewable energy. Aveda continues to increase the use of LED lighting in retail locations even though potential financial savings may accrue to the other parties where utility costs are included in rent. That Aveda does so anyway is an important manifestation of its mission-driven commitment to environmental leadership and responsibility.

7.3.5 Other Dimensions of Aveda's Energy Management – Shipping and Product Use

For a single bottle of shampoo, the energy and carbon footprint represented by inbound shipping of raw materials and packaging and outbound shipping of finished product is relatively minor. Aggregated shipping of all raw materials, packaging, and finished product results in a large amount of energy use and associated carbon footprint. To reduce use and emissions, Aveda maximizes sea and rail transport wherever and whenever feasible. Raw materials are ordered sufficiently far in advance to allow the use of slower, more energy efficient shipping modes, although this is not always practical. Air shipment – both inbound and outbound – is avoided, but truck transportation, especially for outbound product in North America where a significant percentage of Aveda's business occurs, is commonly used. For this mode, the Aveda logistics teams works closely with third-party shippers to maximize loads, minimize route distances and encourage that the most efficient vehicles available are used. As a member of the US EPA's SmartWay program, Aveda works closely with shipping companies on energy and carbon efficiency issues. Aveda has been recognized as a SmartWay shipper since 2006.

Beyond modal changes, Aveda is rethinking what it ships. Shampoo and our other personal care products are a mixture of ingredients that provide functional benefits for hair and skin. To deliver these benefits, the ingredients are mixed with water and put into a package; neither the water, nor the package provide functionality, but they are necessary as delivery mechanisms. Reducing the volume and weight of packaging as a portion of the product–package unit is part of Aveda's broader product packaging strategy (discussed in more detail in the next section). Examples include simplifying packaging to reduce the volume used per product unit, replacing heavier (and less safe in use) glass with lighter plastic, and reducing the overall packaging-to-product ratio by promoting bulk packaging. These innovations typically result in both environmental and financial gains, by reducing costs per unit and increasing shipping efficiency by reducing overall packaging weight. Aveda's Research and Development team is actively seeking ways to reduce the amount of water in product that needs to be shipped to end-users of our products.

Finally, Aveda strives to encourage energy efficiency in product use. This is complicated due to consumer preferences for an overall product-use experience that includes such elements as the pleasure associated with a warm shower. While many of Aveda's products are designed to efficiently achieve efficacy for consumers, Aveda has not yet figured out a marketing strategy to effectively promote cold showers or baths, for example. Thus while the products are formulated to work efficiently with ambient temperature water, they are unlikely to do so by virtue of the way that people choose to use them.

In contrast, a high proportion of Aveda's products are used by independent business partners in Aveda's network of over 8000 salons and spas. Aveda has long worked with this network to promote energy efficiency as a part of broader environmental stewardship program that strengthens their alignment with the Aveda brand. While Aveda cannot *tell* salon/spa owners how to run their business, they are encouraged to align with the brand and Aveda provides many resources to guide them in doing so. As early as 1992, Aveda had a program in place called the "Earth Action Recycling Program," where Aveda field sales people worked with salons to promote energy efficiency as well as other environmental practices!

Other partnership examples include tip sheets, links to local resources for renewable energy, forums for owners to exchange experiences, and, most recently, the creation of an interactive web site for owners with comprehensive and easy to follow guidance and resources to design and implement "the sustainable salon of the future." Energy efficiency is a major focus of this web site, and links for resources ranging from lighting equipment to low-flow water systems to solar energy systems and architects who can help design and implement all of these are provided. Many salon owners have already implemented efficient practices and technologies and it is Aveda's hope that by providing ever-more-accessible tools and resources the many achievements that the brand has made can be leveraged much more widely through the salon-spa network.

7.4 WASTE MANAGEMENT AT AVEDA

In the world of cradle-to-cradle thinking, waste does not exist – only materials. Long before C2C became a widely known concept in the world of materials management, Aveda adopted an approach to materials that aligned well with this concept. Initially focused on materials used in manufacturing, distribution, and office operations, the approach was extended rapidly to Aveda products.

7.4.1 Waste Management in Operations

In its earliest operations, before manufacturing began at the Blaine facility, and, according to early employees, before Aveda signed the Valdez Principles and established formal programs to address waste, Aveda employees were already recycling glass, paper, cardboard, packaging, and other materials locally in Minnesota. With the signing of the Principles and the subsequent move to the Blaine facility, organized programs were created and tracking of waste began. Inspiration seems to have come from an early Aveda "Ecological Mission Statement (undated)" with the following statement:

> We advocate a resourceful society committed to the reduction of materials, reuse of by-products, recycling of spent resources and the purchase of materials made of renewable, recycled and recyclable materials.

A "no waste" goal was established in 1991, a baseline established, and incremental goals for various materials were adopted. In the following year, the variety of materials recycled included three colors of glass, high-density polyethylene (HDPE) and low-density polyethylene (LDPE) plastic (over 4 tons), steel, plastic, and cardboard drums, as well as nearly 50 tons of cardboard. The average recycled content rate for unit cartons and corrugated cardboard was 50%. The program has been steadily expanded over the nearly 20 years of operations in Blaine.

In 1992 and 1993, additional records are available that show in great detail the range of the recycling and reuse programs that were in place and the diverse materials that were addressed, ranging from office paper to shrinkwrap in manufacturing and distribution, to ribbons, ink cartridges, and toner cartridges in the management information systems (MIS) department.

Today this comprehensive approach to recycling and reuse of materials has continued and been extended to include electronic equipment and other difficult-to-recycle materials. Recycling stations are distributed widely throughout Aveda facilities – especially in office areas and on the manufacturing floor. Waste containers are placed such that they are much harder to find, subtly encouraging employees to recycle everything possible (Figure 7.4).

Aveda not only recycles materials used in business operations but encourages employees to bring in materials that are not easily recyclable where they live (e.g., batteries, compact fluorescent bulbs). The net result of these programs is that 85%

Waste Stream	2008	2011	Percent Change
Municipal Waste	479,670	270,050	−44
Hazardous Waste	35,597	21,188	−40
Industrial Waste (product)	242,970	167,359	−31
Corrugate – Recycled	742,430	886,700	19
Recycled and Reused	415,812	286,378	−31

Figure 7.4 Waste Disposal, Recycling and Reuse at Aveda's Primary Manufacturing and Distribution Site.

*quantities in pounds

Source: Reprinted with permission from © Aveda.

of the materials in the manufacturing facility and 99% in Aveda exclusive distribution facilities are either reused (e.g., some shipping cartons are cycled back and forth from packaging suppliers multiple times and a number of suppliers will refill ingredient totes with the same ingredient), recycled, or composted. The remainder is sent to municipal waste-to-energy facilities, so that Aveda has become a no-waste-to-landfill operation. Aveda has recently begun a product delivery system in which reusable plastic totes are used in place of corrugate boxes. While working well in the Minneapolis metropolitan area with some customers, Aveda has not yet figured out how to scale this concept more broadly.

7.4.2 Recycling beyond Blaine

Recycling in Aveda's stores and field offices is more difficult than at headquarters because many shopping malls – where most stores are located – and commercial offices have very limited recycling facilities. Product returns, samplers, and outdated product and packaging are returned to Minnesota for appropriate handling; where local recycling is limited, staff will include additional recyclable materials in the shipping containers. Frequently, Aveda employees in retail locations take it upon themselves to take materials home for recycling – a striking manifestation of the deep commitment that many employees have to the Aveda mission and brand.

7.4.3 Products and Packaging

A cosmetic product (e.g., liquid shampoo) and its associated package (the bottle) are considered a single unit and most sales reflect product–package combinations, which, for Aveda, drive its approach to both product and package design and management.

7.4.3.1 Products

The life cycle of most Aveda products (the exceptions are some accessories such as hair brushes) include plant-based sources for ingredients that are formulated for functional uses designed to be applied, washed off, and returned to the environment through wastewater streams. In its mission driven efforts, Aveda has sought to address the environmental impacts of each stage of the product life cycle; for example, by committing to increasing the use of certified organic ingredients that reduce environmental harms associated with conventional agricultural inputs and contribute to the well-being of farmers. Similarly, Aveda has embarked on a path to create products that can be safely incorporated into natural cycles without causing harm. In C2C terminology, products – assemblages of diverse ingredients – that

Product	Category	Certification Level – Product	Certification Level – Packaging
Dry Remedy	Shampoo	Gold	Silver
Dry Remedy	Conditioner	Gold	Silver
Men's	Shampoo	Gold	Silver
Men's	Conditioner	Gold	Silver
Invati	Shampoo	Gold	Silver
Invati	Conditioner	Gold	Silver
Invati	Revitalizer	Silver	Silver

Figure 7.5 Aveda's Cradle-to-Cradle® Products and Packaging.
Source: Reprinted with permission from © Aveda.

can be safely returned to the environment become biological nutrients and can be continuously cycled. This approach to product development fits perfectly with Aveda's long-held product design philosophy of using plant-derived ingredients. As a way of demonstrating this commitment in the marketplace, Aveda has submitted several products for C2C certification in recent years (see Figure 7.5).

Inherent in the creation and marketing of cosmetic products are waste challenges that Aveda, as well as its peers in the cosmetics industry, must address. Certain wastes, such as pilot batches produced in the research and development process, are planned and unavoidable. Others that occur in the manufacturing process include routine residual product in mix tanks, transfer lines, and fillers that must be cleaned and removed prior to subsequent batches as well as occasional batches that do not turn out as planned. Aveda has an agreement to send such wastes at permitted concentrations to its municipal waste water treatment facility. While Aveda always seeks to minimize such wastes, the nature of the business is such that these do occur and must be managed responsibly.

Additional waste occurs when manufactured product is not sold within specified time frames, typically because sales forecasts do not align with actual sales results. Product that falls into this category is either donated to charitable organizations if it still meets safety standards or disposed of through a licensed waste handling company. Reducing the potential for such wastes is important as they are costly both to manufacture and store and to dispose of properly. Consequently, Aveda is working to improve sales projections via "ground up" forecasts by field sales personnel and to increase product shelf life through improved formulations.

7.4.3.2 Packaging

Cosmetic packaging is multi-functional. While its primary purpose is to contain the product for physical delivery from production point to end user, it must also provide protection from contamination and deterioration, thus keeping it safe for the user, communicating a large amount of information that addresses consumer interests and regulatory requirements, and helping sell the product. After it does all this, it becomes waste if not managed effectively. In this situation, Aveda's design approach is to minimize waste across the packaging system by:

- Using the least amount of packaging necessary to meet design criteria.
- Using materials that can be recycled through public recycling streams wherever feasible.
- Using the highest post-consumer recycled-content (PCR) materials available whenever possible to promote recycling. And
- Working closely with packaging suppliers to assure that Aveda packaging is produced in an environmentally responsible way (e.g., encouraging the use of renewable energy in packaging manufacture).

As cited earlier in this chapter, Aveda has been committed to improving product packaging across its entire product line since the foundation of the brand. Growth through the 1990s gave Aveda the opportunity to leverage key suppliers in conjunction with innovations in environmental packaging technology and establish a leadership role on packaging in the cosmetics industry.

Materials Used

Due to the need for packaging that is compatible with product formulae, is lightweight (for shipping efficiency and ease of use), and is safe in wet environments, Aveda's primary packaging materials are polymer plastics and the focus from the outset has been to use recyclable resins with the highest possible post-consumer recycled (PCR) content. Shampoos and conditioners are among Aveda's highest volume products and these have been packaged in HDPE (high-density polyethylene) bottles with polypropylene (PP) caps from the earliest days. In 1993, Aveda introduced HDPE bottles with 25% PCR content, achieving an industry first. A commitment to increasing PCR content has brought Aveda to the point where its HDPE containers now average 80% PCR content and some use up to 100% PCR. In tubes, another widely-used packaging type, Aveda pioneered PCR content, currently achieving up to 65% PCR and expecting to introduce even higher PCR content tubes. Other products are packaged in up to 100% PET bottles and jars.

By July 2010, 55% of all Aveda packaging was made with recycled content (much of it post-consumer recycled), equivalent to nearly 2 million kilograms of material. Aveda's packaging suppliers estimated that in 2007, Aveda's use of post-consumer recycled plastics resulted in over 450 000 kilograms (1 million pounds) of plastics not going to landfills or incineration.

In addition to plastics, Aveda uses recycled glass containers and, in some candle products, repurposed glass from beverage bottles. Paperboard packaging and printed marketing collateral incorporate up to 100% PCR content in most cases. Aveda has also used non-solvent, agri-based printing inks since 1994. Other industry leading initiatives include the introduction of candle refills to the industry and pioneering work in refillable lipstick cases made from high recycled content resins.

Material Recovery

Aveda recognizes that a commitment to using environmentally preferable packaging materials alone is not going to achieve the desired goal of continuous cycling of materials – thus eliminating the concept of packaging as waste – unless an effective "closed-loop" system for materials recovery exists. While the European Packaging Directive, the "Green-Dot" program requiring manufacturers to take responsibility for packaging materials introduced to the marketplace, is an important step in the direction of continuous cycling of materials, it does not fully achieve the goal since some "recovered" materials are not recycled. Elsewhere in the world even less systematic approaches exist and much packaging material ends up as waste no matter whether public recycling capacity exists or not.

If the cradle-to-cradle vision for eliminating the concept of waste is to be realized, systems must be in place to collect and repurpose materials that would otherwise become waste. Aveda has sought to create markets for recycled plastics, paperboard and glass, for example, by "pushing the envelope" to utilize these materials in its packaging. However, Aveda fully realizes that this is only a step toward the ultimate goal and has initiated experiments in the past few years to establish such systems.

In 2008 Aveda launched the "Recycle Caps with Aveda" initiative to promote the recycling of PP resin. This was done:

- To encourage collection and repurposing (Aveda uses 100% PCR content caps on some products) of a resin not commonly collected in public recycling programs in the USA.
- To increase awareness of the importance of recycling among youth by structuring the program so that the primary collectors are school children.
- To encourage municipalities to add this resin to their programs by demonstrating that there is a market for the material.

The program has generated participation by hundreds of schools and other organizations that have retrieved 230 tons of polypropylene (approaching 100 million caps) that have been recycled into caps and other products and has resulted in several US municipalities adding polypropylene recycling to their collections in response to "lobbying" by Aveda program participants in their communities. Although the efforts are worthwhile, scaling the program to capture a high proportion of the PP resin currently in the wastestream is beyond Aveda's limited resources and the company is cooperating in efforts to expand collection and recycling.

Beyond caps, Aveda recognized that an approach to continuously retrieving and repurposing packaging materials requires a focused process and, in 2010, began a pilot program in one US sales region. The program tested the feasibility of retrieving all packaging materials that are not readily recyclable in public programs through its partner salons and retail locations, and engaged a third-party partner to find end uses for the collected material. While still in the testing phase at the time of writing, the preliminary results indicate that over 98% of the materials collected can be usefully repurposed either for new packaging materials or for other uses. Aveda is currently evaluating the logistical and financial aspects of expanding such a process more widely.

Aveda initiated both the caps program and the pilot collection after realizing that, no matter how innovative and aligned with its mission and sustainability policy its approach to packaging development has been, a large proportion (exact number unknown) of its packaging is not recovered and repurposed, despite extensive public programs to encourage and facilitate recycling in various parts of the world.

7.5 SUMMARY

Energy and waste management in the cosmetics industry presents complex challenges. Products utilize many and diverse ingredients from numerous sources around the globe; manufacturing processes can be complex and lengthy; and product packaging is often substantial for a variety of reasons ranging from product safety to marketing objectives. By taking a systematic, life-cycle based, approach to innovation in energy use and materials management, cosmetics companies have the potential to make a substantial contribution to meeting growing global sustainability challenges resulting from increasingly scarce resources.

For energy management, opportunities range from improving the efficiency in the use of current energy resources throughout a product's life cycle to – ultimately – fully integrating renewable energy throughout a business. For waste, the opportunity is growing to eliminate the concept and create a world in which materials can be continuously cycled as "food" or "nutrients" throughout the global industrial system. Aveda has strived from the founding of its business – and continues to strive today – to integrate these visionary energy and materials management principles into all aspects of business; and welcomes the increased commitment to doing so – and the knowledge that results – across the industry.

REFERENCES

[1] Richman, S. Our Ultimate Resource Gone, The Future of Freedom Foundation; http://www.fff.org/comment/ed0298d.asp (accessed on July 9, 2012).

[2] Tierney, J. (1990) Betting on the Planet, *New York* Times, December 2, 1990. http://www.nytimes.com/1990/12/02/magazine/betting-on-the-planet.html?pagewanted=all&src=pm (accessed on July 9, 2012).
[3] Cordell, D., Drangert, J-O. and White, S. (2009) The story of phosphorus: global food security and food for thought. *Global Environmental Change*, **19**, 292–305.
[4] United States Environmental Protection Agency. http://water.epa.gov/polwaste/nps/agriculture.cfm (accessed on July 9, 2012).

8

Corporate Social Responsibility and Philanthropy

PART 1: INTRODUCTION

Amarjit Sahota

8.1.1 CORPORATE SOCIAL RESPONSIBILITY

The Corporate Social Responsibility (CSR) term has many interpretations that vary from country to country. Developing countries tend to place more emphasis on product quality, whereas industrialized countries define CSR according to labor issues, the environment, and community involvement.

Ideally, CSR policy would function as a built-in, self-regulating mechanism, whereby organizations would monitor and ensure their adherence to law, ethical standards, and international norms. Organizations would take responsibility for the impact of their activities on the environment, consumers, employees, social communities, stakeholders, and other members of the public. Furthermore, organizations would proactively promote public interests by encouraging community growth and development, and voluntarily eliminate practices that harm the public, regardless of legality. Essentially, CSR is the deliberate inclusion of public interest into corporate decision-making, and the honoring of a triple bottom line: People, Planet, Profits [1].

The European Commission defines CSR as...

> A concept whereby companies integrate social and environmental concerns in their business operations and in their interaction with their stakeholders on a voluntary basis

Sustainability: How the Cosmetics Industry is Greening up, First Edition. Edited by Amarjit Sahota.

Among other things, this definition helps to emphasize that:

- CSR covers social *and* environmental issues, in spite of the English term corporate *social* responsibility.
- CSR is not or should not be separate from business strategy and operations: it is about integrating social and environmental concerns into business strategy and operations.
- CSR is a voluntary concept.

An important aspect of CSR is how organizations interact with their internal and external stakeholders. Such stakeholder groups include employees, customers, neighbors, non-governmental organizations, public authorities, and so on.

Although the terms are similar, CSR and sustainability are not the same. A detailed explanation of the sustainability term is given in Chapter 1 (Section 1.2). The major differences between CSR and sustainability are best illustrated in Figure 8.1.1 [2].

The premise of CSR is that the success of an organization is dependent on people (society) and resources (the environment), so an organization should take responsibility for its influences on society and the environment. The success of an organization is therefore interdependent on the success of society and the environment.

Sustainability has a wider remit in that it takes into account the needs of today and the needs of future generations, whilst working with resource limitations.

8.1.2 CORPORATE PHILANTHROPY

Corporate philanthropy refers to the act of giving a portion of profits or resources to good causes; these good causes can be to the arts, education, housing, health, social welfare, and the environment, among others, but excludes political contributions and commercial sponsorship.

Businesses see corporate philanthropy as a way for them to share benefits with the communities and environment they operate in. Large corporations across industries have long been involved in corporate philanthropy. For instance, many large American companies set up private foundations for "good causes" in the twentieth century; examples include the Rockefeller, Carnegie, and Ford foundations. The practice continues today, with a number of cosmetic companies setting up foundations.

The Avon Foundation was set up by the American cosmetics company in 1955. It is now the world's largest corporate-affiliated philanthropy that focuses on women's issues. Two of its major focuses are breast cancer research and violence against

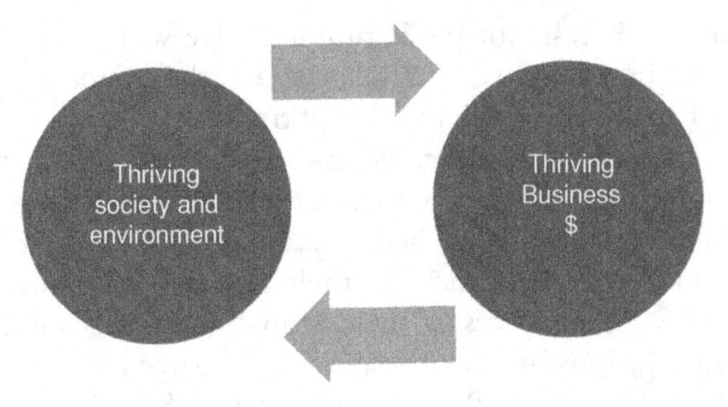

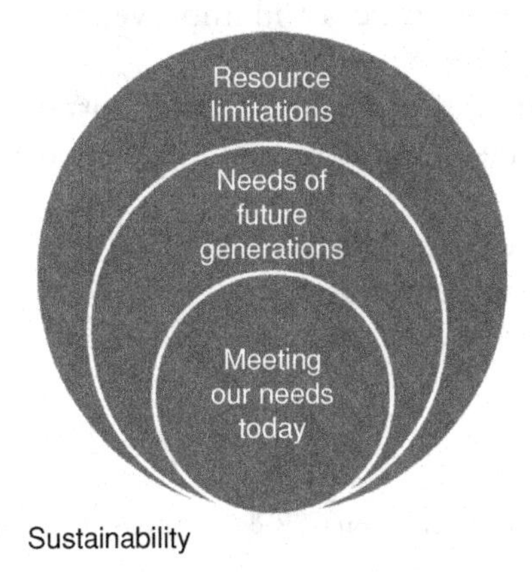

Figure 8.1.1 Difference between CSR and Sustainability.
Source: Reprinted with permission from Kevin Moss at www.csrperspective.com.

women. It is also provides aid to natural disasters and emergency programs. Avon has donated almost US $1 billion to the foundation since its formation.

The Brazilian cosmetics company, Grupo Boticário, set up a private foundation for nature conservation in 1990. Fundação Grupo Boticário promotes and undertakes nature conservation projects in Brazil. The foundation has set up 1306 conservation projects leading to the discovery of 43 new species [3].

A more common practice is to donate a portion of profits or sales to charitable causes. About 40 cosmetic companies are members of the US-based 1% for the Planet (1% FTP) organization. These companies donate 1% of their sales to environmental organizations. In return, the companies can place the 1% FTP member logos on their corporate stationary.

Some cosmetic companies, like Éminence Organic Skin Care, are taking more direct action. Starting in November 2012, the Canadian company started donating proceeds from the sale of its active organic or biodynamic products to combat climate change. The Forests for the Future initiative will plant one tree for every such product sold. In the first year of the program, the company estimates to plant the equivalent of over 1000 football fields of trees.

As society's expectations of companies rise and as many companies start operating in more distant locations, they are expected to address a growing list of needs. Companies that 20 years ago were held accountable only for direct, contractually specified, or regulated consequences of their actions today find themselves held to account for the consequences of their actions in areas such as environmental sustainability and the governance of resource-rich, low-income nations.

Such realizations have led La Prairie to give about US $1.30 to the Ocean Futures Society for each sale of its Advanced Marine Biology Cream, and Aveda to raise US $2.1 million for protection of water resources and improved access to clean water in developing countries.

Growing consumer expectations of companies make corporate philanthropy more important than ever; it can be an effective tool for companies trying to meet such rising expectations.

The proceeding two case studies demonstrate some best practices in CSR and corporate philanthropy in the cosmetics industry.

REFERENCES

[1] Organic Monitor (2010) Strategic Insights report on CSR & Sustainability in the Cosmetics Industry, London, UK.
[2] Kevin Moss at CSR Perspective http://csrperspective.com.
[3] Sustainable Cosmetics Summit Latin America, Sao Paulo, September 24–26 2012.

PART 2: BURT'S BEES® CASE STUDY[1]

Paula Alexander

Burt's humble beginnings as a low-impact beekeeper illustrate the mutual respect for nature, humans, plants, and animals which continues to steer the BURT'S BEES® business' social and environmental vision for the future. The story's origin sheds light on how the brand remains as committed to corporate social responsibility (CSR) today as it was at its founding. In 1984, Roxanne Quimby, a struggling artist and waitress, met Burt Shavitz, a photojournalist veteran living "off the grid" in Maine. The early days tell a tale of survival for this unexpected duo. Burt was trying to find a use for the leftover beeswax from the hives of his honey

[1]BURT'S BEES and the BURT'S BEES logos are registered trademarks of BURT'S BEES, Inc. Used with permission.

making business, and Roxanne suggested making beeswax candles from the extra material. When they made US $200 on their first day at a craft fair, they knew they were on to something. They eventually stumbled upon a farmer's almanac with a recipe for lip balm. From this time-tested recipe, BURT'S BEES® Beeswax Lip Balm was born.

With Burt attending the first Earth Day in 1970, and Roxanne pursuing conservation projects, both proved themselves eager students of nature and ardent environmental advocates. Together they believed, "because we take from nature we need to protect and preserve nature." From the beginning, their frugality, by necessity, combined with their nature loving spirit, imprinted sustainability into the BURT'S BEES® brand DNA (Figure 8.2.1).

BURT'S BEES®, the brand stamped with a wood-cut image of the bearded man's face on a yellow background, grew for the next decade. The iconic yellow now used on all BURT'S BEES® product labels was inspired by Burt's yellow truck, which he used to draw attention to his products at the farmers market. From selling at the farmers market to entering mass retail channels, the 1990s and 2000s meant major growth and flux for BURT'S BEES®. This internal change coupled with significant market demand from consumers for businesses to be more environmentally and socially responsible, required leadership to think strategically about their CSR approach in the new millennia. The business added new products, moved away from candles to focus exclusively on health and beauty care products, and moved to North Carolina to expand manufacturing and distribution capabilities. Not only did the company experience significant leadership and ownership changes, but

Figure 8.2.1 Roxanne and Burt circa 1990.
Source: © 2013 BURT'S BEES®, Inc. Reprinted with permission.

Roxanne sold a majority share of the company to a private equity firm in 2003. From 2000 to 2007, BURT'S BEES®' annual revenue soared to $164 million from $23 million. BURT'S BEES® was acquired by The Clorox Company in 2007, as part of their Centennial Strategy to continue to address health and wellness and sustainability trends. Throughout the business integration process, Clorox has been committed to ensuring that the BURT'S BEES® business retains the culture and values that make the brand special. To this day, the BURT'S BEES® business continues to seek ways to raise the level of natural ingredients in its formulations, drive footprint reduction, and improve the level of sustainability in its packaging.

8.2.1 VALUE-DRIVEN SUSTAINABILITY LEADERSHIP

Woven through the exponential expansion for the business and the natural health and personal care category, BURT'S BEES® mission today remains consistent from the company's modest origins, "We make people's lives better every day, naturally." To accomplish this mission and bring together two different cultures, those before and after the private equity purchase and expansion, BURT'S BEES® leaders worked with a group of 40 passionate employees to define the cultural values and behaviors, coining the We Care Way. The four values around which the culture was created include Steadfast Commitment to the Greater Good, Harmonious Tension, Embrace Change, and Passionate Teamwork (see Figure 8.2.2). Not only have these

Figure 8.2.2 BURT'S BEES® culture and values.
Source: © 2013 BURT'S BEES®, Inc. Reprinted with permission.

values kept BURT'S BEES® economically viable, but the programs which activate them differentiate BURT'S BEES® from other brands.

The values set expectations for employee behavior and were socialized into the business via Launch It and Live It Teams. The ongoing keepers of the company culture were a group of volunteers called ECOBEES, a concept now embraced by all employees. ECOBEES evolved into The Greater Good teams, and these champions of culture are responsible for education, onboarding, team building, the 100% employee engagement program, and culture events.

8.2.2 THE GREATER GOOD BUSINESS MODEL: AN INTEGRATED APPROACH TO SUSTAINABILITY

The Greater Good Business Model (see Figure 8.2.3) was created to be the compass for the brand's triple-bottom line approach by focusing on the three pillars of natural well-being, sustainability, and social responsibility. Environmental best practices became further institutionalized through formal roles, including a director of sustainability who established the brand's biennial Social and Environmental

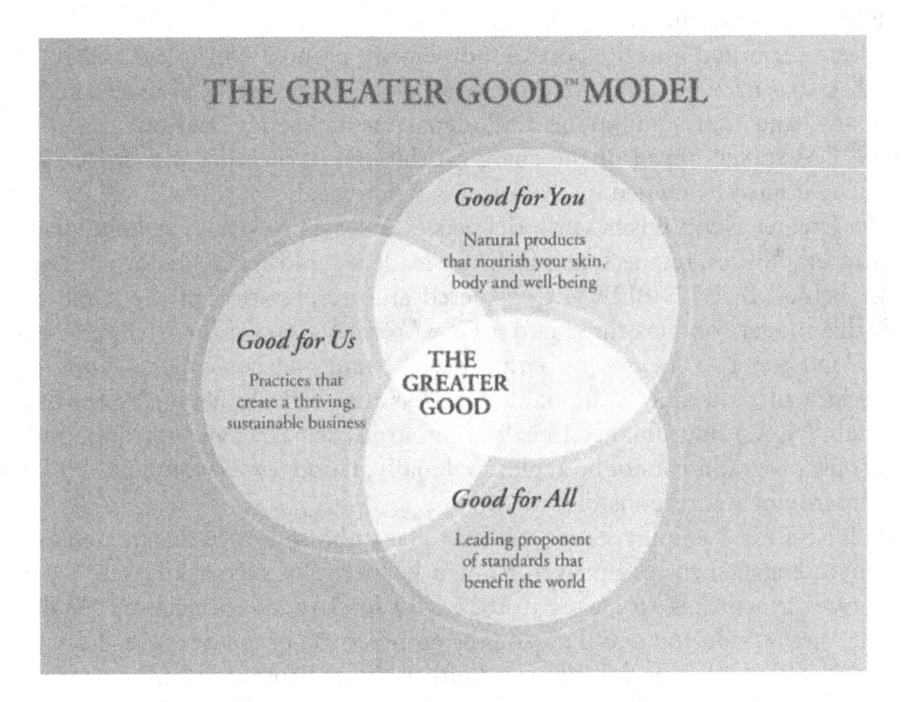

Figure 8.2.3 The Greater Good Business Model.
Source: © 2013 BURT'S BEES®, Inc. Reprinted with permission.

Reports, The Greater Good Foundation as the brand's formal philanthropic arm, and the 2020 Sustainability Goals, which include:

- Product and Packaging – 100% Natural/100% Post-Consumer Recycled/Biodegradable (in progress).
- Waste Reduction – Zero Waste to Landfill (completed).
- Carbon Footprint – 100% Renewable Energy (being evaluated).
- Green Facilities – LEED Platinum/ISO 14001 certified (being evaluated).
- Employee Engagement – 100% Employee Engagement (completed).

The sustainability team realized these goals were not simply a finish line to be crossed, but rather a journey, and renewed strategic priorities required that they were updated in 2012.

As part of its new sustainability strategy, and taking a cue from nature, BURT'S BEES® has taken a whole-systems thinking approach to CSR and its organizational structure. According to Path Tree Consulting, "Whole-systems thinking is a method of analysis and decision-making that looks at the interrelationships of the constituent parts of a system rather than narrowly focusing on the parts themselves"[1]. This improves performance in the long run and reduces blind spots. Implementation of sustainability practices throughout the business was a focus for BURT'S BEES® in the 2000s and has become even more critical as the company has been integrated into the portfolio of Clorox brands. As Unilever CMO, Keith Weed, stated in 2012, "Brands should weave CSR into their practices across the company, and that stand-alone CSR departments are an 'add-on" [2]. BURT'S BEES® has stayed ahead of the curve on this trend, showing that for CSR to be effective, it must be owned by all business functions.

The Greater Good Business Model was established to ensure stakeholders, consumers, employees, retailers, and industry leaders could articulate the guiding principles behind BURT'S BEES®. Considered an interpretation of the triple bottom line, this model contains three parts: *Good for You*, *Good for Us*, and *Good for All*. *Good for You* focuses on natural well-being and consumers. *Good for Us* is the idea of a thriving, sustainable business that actively demonstrates financial profitability. Without financial health, ambitious social, environmental, and philanthropic programs cannot be achieved. Finally, *Good for All* combines both social and environmental responsibility.

BURT'S BEES® employees are trained to be ambassadors of this model, as direct employee engagement has proven to be a key tenet of successful, purpose-driven businesses. According to Hayes Roth, CMO for Landor Associates, "Make sure all employees understand and personally embrace the brand purpose. They are the brand, and if employees don't buy it, neither will the customer" [3].

BURT'S BEES® prides itself on its honesty, authenticity, and transparency – traits that are not only relevant to today's consumer, but which are perpetual throughout the brand's history. To give transparency to consumers, the company states its percentage of natural ingredients in all its products. BURT'S BEES® products are above 95% natural, on average 99% natural. Its products do not include sodium lauryl sulfate (SLS), parabens, phthalates, or petro-chemicals. BURT'S BEES® hosts beauty editors, such as Jean Godfrey June of *Lucky*, at its manufacturing and R&D labs to illustrate this transparency in action. June subsequently reported, "If you want to buy something earnestly, totally eco, buy a BURT'S BEES® item. And if you want to live an earnestly, totally eco life, I suggest moving to Durham, North Carolina, state. It took all my self-restraint to keep from diving into the vat [of Milk & Honey Body Lotion]. With every kind of super-moisturizing oil and butter, plus food-grade emulsifiers, plus of course actual milk and actual honey, you do want to eat the stuff, and you do want to bathe in it" [4].

In an effort to promote transparency across the entire health and beauty care category and to clear up consumer confusion regarding what does and does not constitute natural, BURT'S BEES® worked with the Natural Products Association (NPA), suppliers and competitors to establish the Natural Standard for Personal Care Products in the United States in 2008. While consumer confusion concerning the concept of natural continues to challenge the industry, consumers have become more educated about the ingredients in their personal care products.

BURT'S BEES® launched a print advertising campaign, a promotion with spokesperson Julianne Moore, and a New York City press event on behalf of the NPA. These efforts demonstrated how the company's size, scale, and high consumer awareness could help make natural personal care more available to consumers seeking a healthier lifestyle. The company focused its natural education initiatives on the retail environment and used in-store real estate to promote the NPA Standard, which remains the most widely recognized third-party personal care certification in the United States. BURT'S BEES® engages with NaTrue, NSF, and COSMOS to drive international multi-stakeholder guidance and dialog. Consumer awareness and understanding of 'natural' not only protects BURT'S BEES®' interests, but helps distinguish and grow the category.

A commitment to high CSR standards and transparency requires understanding and work across the entire value chain. Whole-systems thinking has led to a 360° approach to sustainability at BURT'S BEES®, thereby affecting natural products, responsible sourcing, packaging, operations, information systems, workplace practices, and outreach. BURT'S BEES® became zero waste to landfill in 2010 through educational activities, like dumpster diving, as well as a commitment to finding new recycling streams, compost providers, and overall waste reduction mechanisms. On the packaging front, BURT'S BEES® has innovated new materials, such as wrappers made from crushed stone and trays made from potato starch. The post-consumer recycled content which goes into plastic and cardboard packaging components has increased significantly, earning the company several awards in the industry. Strategic sustainability priorities are focused on responsible sourcing, innovative packaging, footprint reduction, and outreach. These categories represent steps along the value chain from source to disposal that can make the greatest positive impact on The Greater Good.

8.2.3 STRATEGIC GIVING

With a strong commitment from company leadership to support the local community and state of North Carolina where the business is based, the Greater Good Foundation was established in 2007 to serve as the philanthropic arm of BURT'S BEES®. The Foundation is funded by 10% of burtsbees.com sales, and the business remains dedicated to contributing a fixed percentage of sales. Foundation funding will only continue to grow as online sales at burtsbees.com, and the personal care category at large, increase.

The Foundation provides two types of giving: (1) Strategic: programs that deliver triple bottom line benefits and support The Greater Good and (2) Good Neighbor: programs that support being a good neighbor in the Durham, North Carolina community. In 2011, 37% of funding supported environmental programs, 39% backed social programs, and 17% met triple-bottom line criteria (Figure 8.2.4). Foundation giving is not to be used to benefit the company financially. However, the more strategically the company chooses issues that support The Greater Good Model criteria, the more likely that a program will deliver "sufficient social value for the company and stakeholders to justify the resources used in creating that value," according to University of North Carolina Kenan Flagler Business School Professor, Lisa Jones Christensen, Ph.D [5]. BURT'S BEES® sees this social value

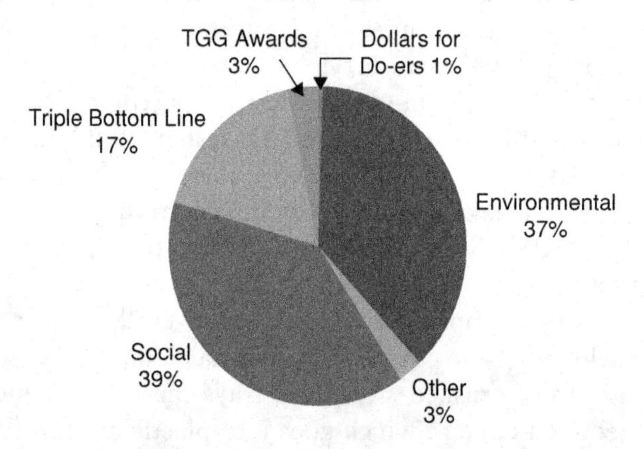

Figure 8.2.4 The Greater Good Foundation 2011 Giving Foundation.
Source: © 2013 BURT'S BEES®, Inc. Reprinted with permission.

manifest itself in terms of employee retention, customer loyalty, and positive brand reputation.

BURT'S BEES® annually recognizes employees who have demonstrated the brand's values in the community through The Greater Good Awards, including the Burt Shavitz Environmental Stewardship Award, the Roxanne Quimby Humanitarian Award, the Renee Quimby Growth Leader Award, and the BURT'S BEES® Lifestyle Change Award. Each winning recipient nominated by his or her peers receives a cash reward, a matching contribution to a charity of his or her choice, as well as a hand carved wood spirit, a symbol representing employees' efforts to protect The Greater Good Values. The award-winning employee contributions to non-profits of their choice represented 3% of the Foundation giving in 2011. BURT'S BEES® also recognizes employees who contribute significant volunteer hours with local non-profits in the Dollars for Do-ers Program.

Examples of Good Neighbor giving from BURT'S BEES® are donations to Habitat for Humanity, Teach for America, and the North Carolina Museum of Natural Sciences. Not only has the company contributed funding for these programs, it has also averaged 5000 employee service hours per year. Company leaders serve on the board of Habitat for Humanity of Durham, and staff has helped to build five homes, one of which includes North Carolina's first LEED certified low–moderate income community.

Since 2009, BURT'S BEES® has teamed up with Teach for America, sending executives into the classroom for Teach for America Week to share their knowledge and be role models of what is possible through education. BURT'S BEES® "Talked Trash" with students to teach them about waste and the four Rs: Reduce, Reuse, Recycle, and Rot (composting). In 2012, using the Pollinator Partnership's Bee Smart Curriculum, BURT'S BEES® taught about the importance of pollinators, Colony Collapse Disorder, and what students can do to help.

The business partnered with the North Carolina Museum of Natural Sciences (NCMNS) to establish an annual Earth Day event called Planet Earth Celebration to raise awareness and enthusiasm about environmental issues and solutions. The goal was to create a venue and platform for non-profit agencies to meet and interact with like-minded groups they might not otherwise meet. BURT'S BEES® has supported the event with sponsorship, planning resources, volunteers, and a tent sale that attracted thousands of people. All proceeds from the 2012 tent sale benefited Triangle Residential Options for Substance Abusers (TROSA), a local non-profit. This event continues to have a green focus on recycling, composting, and responsible use of resources.

For many non-profits, volunteer hours, in-kind services, product donations, and advocacy support can be as valuable as monetary contributions. Product donations have included personal hygiene items for groups in developing countries through AmeriCares and local groups like TROSA. BURT'S BEES® employees volunteer their time at a number of non-profit organizations during business hours. The team supports multiple programs, including Habitat for Humanity house builds, and participates in outreach events and Culture Days. Other companies offer volunteer days, but often work is done in personal time independent of the company. For a corporate outreach program to be successful, it must be enduring and holistic. One-time programs do not support corporate or non-profit goals and can result in wasted time and resources. In addition, BURT'S BEES® stakeholders are likely to support and appreciate sustainable cause programs. If a corporate giving program is based on authentic values, strategic planning, and follow through, the outcomes will illustrate a genuine interest in improving the communities it serves.

8.2.4 EMPLOYEE ENGAGEMENT

Each employee acting as a leader brings The Greater Good Model to life. The high quality talent that BURT'S BEES® looks to attract and retain is due in part to the strong company culture. A healthy company culture must be fostered by executive leadership, especially in a climate of continuous change and fast-paced growth such as that which BURT'S BEES® has experienced. The business sustains this culture through annual Culture Day events and the Live the Greater Good Program (LTGG).

Since its 2009 inception, LTGG has succeeded in its primary objective of driving 100% employee engagement. The program consists of four modules: *Wellness*, *Environment*, *Outreach*, and *Leadership* and requires eight hours of every employee's time for learning, team building, and community giving. LTGG is led by an all-volunteer group of sponsors who oversee the program, captains who coordinate smaller team participation, and content leaders who have a particular interest or expertise in an area.

The *Wellness Module* has included talks on stress management, eating on the go, and a visit from the Kashi/BURT'S BEES® Wellness Mobile Tour. The *Environment Module* has informed employees about hybrid and electric cars, safe homemade household cleaning, and composting. The *Leadership Module* has offered training on managing change, biomimicry, and innovation, including a popular live honey bee display on honey bee health. The *Outreach* Module provides the greatest community impact and rewarding results for employees via:

- Preparing much needed classroom materials for Teach for America.
- Fixing meals "assembly line style" for the hunger relief agency Stop Hunger Now.
- Building hiking trails and lake clean up at Lake Crabtree near BURT'S BEES (R) distribution center.
- Building and maintaining urban agriculture and community gardens in downtown Durham with Natural Environmental Ecological Management (NEEM).

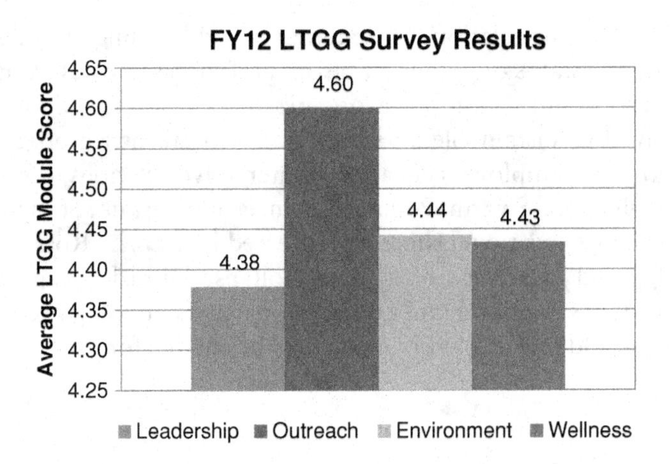

Figure 8.2.5 Live The Greater Good Satisfaction Scores.
Source: © 2013 BURT'S BEES®, Inc. Reprinted with permission.

LTGG continues to be a source of pride for employees, providing a fun atmosphere to collaborate among team members and an important way to contribute to the community. Employees complete a survey after each program that helps content planners continually improve the program. The 2012 average rating on a five point scale was 4.5, an indicator of success and a notable measure of ongoing support from employees. The program will continue to evolve as the company works to address the different employee populations in the manufacturing facilities as well as commercial, sales, and international field offices around the globe (see Figure 8.2.5).

The annual Culture Day event serves as an immersion into the culture for new hires and a recommitment for current employees to the BURT'S BEES® brand values. Each year the event has a different theme that addresses current needs of the company and the surrounding community. Previous years have included an introduction to the We Care Way values and community service projects such as building a children's playground and establishing community gardens.

Culture Day 2012 event kicked off the brand's honeybee health campaign, Wild for Bees™, in conjunction with National Pollinator Week. BURT'S BEES® has partnered with the Pollinator Partnership (PP) and the North American Pollinator Protection Campaign (NAPPC) since 2007 to fund research for honeybee health to combat Colony Collapse Disorder, the mysterious disappearance of honeybees. Honeybees are important because they are an indicator species that provide evidence of issues in the environment. Along with other pollinators like bumblebees, butterflies, and bats, honeybees are responsible for pollinating healthy foods which contribute one third of all food consumed in North America.

Culture Day 2012 embodied BURT'S BEES® integrated CSR program that started internally with strategic consistency and was then communicated through social media and public relations for external brand building. The day began with an overview on the status of pollinator health and success stories from four Greater Good Foundation funded non-profit organizations from The Conservation Fund that are involved in sustainable agriculture and social justice programs throughout North Carolina. Employees engaged in interactive activities linked to company values and Bee-haviors: Swarm, Pollinate, Dance, and Sustain. Staff viewed the premiere of three new educational short films created by Isabella Rossellini to increase bee knowledge and awareness of Colony Collapse Disorder. Employees worked with 21 of The Conservation Fund non-profit partners building bee boxes and seed bomb kits to be sold as fundraisers or used as pollinator forage.

8.2.5 SUMMARY

CSR is not just a program at BURT'S BEES®. Respect for nature and people has been a guiding principle ever since Burt and Roxanne first started making candles from excess beeswax. Sustainability is part of BURT'S BEES® heritage, embedded into all aspects of the brand's culture, operations, planning, and outreach. The Greater Good Business Model is considered in every conversation to ensure business decisions are made balancing the triple-bottom line. BURT'S BEES® believes that business has the power to create positive change. This philosophy and the brand's purpose drives the passion of BURT'S BEES® employees and loyal consumers, directs the brand's strategic giving, creates value for shareholders, and ultimately delivers palpable social impact in all of the communities BURT'S BEES® serves. Perhaps the best way to think of CSR at BURT'S BEES® "is to simply learn from the bee itself. Bees live in harmony, cooperate with each other and share information so the hive benefits. Everything they collect, they turn into something better. Everything they consume they put to wonderful uses. Not a bad role model for a brand that all started with a beekeeper and his bees' wax," states Nick Vlahos, VP and General Manager, BURT'S BEES® [6].

REFERENCES

[1] Whole-Systems Thinking. (n.d.) In Path Tree Whole Systems Thinking web page. Retrieved September 17, 2012 from http://pathtree.com/whole-systems-living.

[2] CSR Departments Redundant, Unilever Chief Says. (2011, November 30). *Environmental Leader*. Retrieved December 12, 2011, from http://www.environmentalleader.com/2011/11/30/csr-departments-redundant-unilever-chief-says/.

[3] Roth, H. (2012). Glass Houses. *The Hub Magazine*, (July/August) 21.

[4] June, J.G. (2010). BURT'S BEES (R) Naturally Nourishing Milk & Honey Body Lotion. *Lucky Magazine* (September) Retrieved August 6, 2012, from http://www.luckymag.com/beauty/2010/09/burts-bees-milk-honey-lotion.print.

[5] Jones Christensen, L. (2012) "Corporate Philanthropy Seminar." North Carolina Network of Grant Makers [Seminar]. Chapel Hill, North Carolina. Mar. 2012.
[6] Burts' Bees (n.d.), Press Kit. Retrieved September 17, 2012 from http://www.burtsbees.com/c/root-press-kit-burt-s-bees.html.

PART 3: DR. BRONNER'S MAGIC SOAPS: BUSINESS AS ACTIVISM

David Bronner

8.3.1 INTRODUCTION

At Dr. Bronner's Magic Soaps, the bottom line is subservient to their activism on behalf of worthwhile causes. However, their activism requires a healthy bottom line. Since 2007, Dr. Bronner's spending on social and environmental causes and charities has roughly matched total after-tax income.

Total compensation of executives is capped at five times that of the lowest-paid position.

Employees annually receive 15% of salary paid into a retirement/profit-sharing plan, up to 25% of salary as a bonus, and a non-deductible preferred provider organization (PPO) health insurance plan for themselves and their families.

The over 30 000 words spread across all the soap labels were Dr. Bronner's life work of searching every religion and philosophy for "Full Truths" that can be summed up in two beautiful sentences:

1. CONSTRUCTIVE CAPITALISM IS WHERE YOU SHARE THE PROFIT WITH THE WORKERS AND THE EARTH FROM WHICH YOU MADE IT!
2. WE ARE ALL BROTHERS AND SISTERS AND WE SHOULD TAKE CARE OF EACH OTHER AND SPACESHIP EARTH!

Major causes and focuses in 2012 include fighting for organic integrity in personal care, recommercialization of cannabis, and promotion of "Fair Trade" certification of product supply chains for companies in the natural food and body care channel.

8.3.2 COMPANY BACKGROUND

Dr. E.H. Bronner, grandfather of the current owners, founded Dr. Bronner's in 1948. He ran the enterprise as a non-profit religious organization, promoting his vision on every bottle of soap that all the great spiritual giants of faith traditions

Figure 8.3.1 Dr. Bronner's Magic Soaps product range.
Source: © Dr. Bronner's Magic Soaps.

are at their heart saying the same thing: we are all children of one divine source, and if we don't get past the trivial ethnic and religious differences that divide us and realize our transcendent unity, we will perish. "We're All-One or None!" is the mantra on their labels (Figure 8.3.1).

Dr. E.H. Bronner was a third-generation master soapmaker from a German–Jewish soapmaking family who emigrated to the USA in 1929. His parents and much of his extended family were killed in the Holocaust. He gave away most of his profits to social and environmental initiatives; however, the Internal Revenue Service (IRS) did not agree with the non-profit status of his enterprise. He lost a legal case in the late 1980s, leaving the company bankrupt and owing huge back tax payments and penalties. The founder's sons – Jim and Ralph Bronner – along with Jim's wife, Trudy Bronner, stepped in and reorganized it as a for-profit enterprise; they got the company out of bankruptcy and on sound financial footing. Jim had set up his own specialty chemical consulting firm, Bronner Chemical, where his son (the contributor David Bronner) grew up working. At Bronner Chemical – among other things – Jim developed the world's leading fire-fighting foam concentrate, and then modified it into a fake snow for Hollywood.

Dr. Bronner died in 1997 and his son Jim died in 1998; the company is now run by Jim's two sons – David and Mike Bronner – grandsons of the company founder. The company remains a for-profit organization but with non-profit DNA. The company is exploring alternative corporate structures to reflect its "social enterprise" mission. However, unlike the company founder Dr. Bronner, the current owners do not sermonize and preach his vision and mission in the organization's business and charitable activities, although they do translate the essential core of his message into how the business operates. All profit not needed for business development, taxes, and debt is donated to worthwhile causes and charities worldwide.

8.3.3 FAIR TRADE PROJECTS

In 2002, Dr. Bronner's was determined to shift its major raw materials to certified organic sources. By 2003, all the company's soaps were certified under the USDAs National Organic Program (NOP), but over the next two years the Bronner family realized that its supply chains were opaque to them: they bought from intermediate brokers and did not know whether the organic farmers, farm workers, and factory workers in their supply chains received fair prices and wages, or whether child or exploited labor made their organic oils.

While the organic movement initially had social criteria regarding pricing, wages, and working conditions, these had been completely dropped from the final NOP regulations. Inspired by fair trade brands, such as Equal Exchange and Guayaki, in 2005 the Bronner family decided to commit the company's full financial clout and staff resources to converting all major raw materials and supply chains to certified fair trade status. These materials included organic coconut, palm, olive, and mint oils; they collectively constituted over 95% by weight of their agricultural volume, everything except water and the alkali needed to saponify the soaps. In effect, this switch would allow them to produce "fair trade" soaps.

First, the company researched the basic tenets of fair trade and resolved to follow them: cut out intermediaries in the purchasing chain; know the farmers and their communities from whom they are buying; build long-term trading relationships; make sure prices for crops and wages are fair and paid promptly; help finance farm inputs such as organic compost; set a floor price that guarantees to cover farmers' cost of production (COP) plus a fair profit should market prices crash below COP; ensure that working conditions in processing operations are safe; follow rules set by the International Labor Organization regarding working hours, gender equity, and the right to collective organization; contribute a fair trade premium for community development, such as for medical equipment, health clinics, school books, water sanitation – whatever the local needs might be; and achieve the participation of all stakeholders.

The company then searched for existing producer projects for its main raw materials that would meet fair trade criteria. The fair trade movement had initially emerged around coffee and cocoa: major export commodities whose producers around the world routinely suffered from major price fluctuations and exploitative trading practices. Only recently had the fair trade concept begun to expand into other commodities. The only fair trade supplier of the raw materials required by Dr. Bronner's was the olive oil producer Canaan Fair Trade in Palestine. The company had organized more than 1000 small olive farmers in the West Bank into village groups in cooperation with the Palestine Fair Trade Association (PFTA), and was able to supply the volumes required.

The Swiss certification organization Institute for Marketecology (IMO) launched its "Fair for Life" standard in 2006. In 2007, Canaan Fair Trade was one of the first groups certified under this new fair trade program.[1] Since then, Dr. Bronner's has purchased most of its olive oil from Canaan, more than 100 metric tonnes per year and growing. Deliveries are pre-financed and support is given to Canaan regarding its expansion, since fair trade is about long-term relationships and mutual benefit. The balance of the organic olive is bought from Israel: a portion of which comes from Sindyanna, an Arab–Israeli fair trade project. Dr. Bronner's youngest sister went to the Ein Gev kibbutz in 1936 when she was 16; the current owners therefore have family links throughout Israel.

Although supply of fair trade olive oil was now secured, Dr. Bronner's had no credible fair trade sources for its other main raw materials. The company thus decided to set up organic and fair trade projects for coconut oil in post-tsunami Sri Lanka, for palm oil in Ghana, and for mint oil in India. In Spring 2007, after organizing fair trade organic coconut farmers and investing close to USD 2 million in setting up a modern factory for virgin coconut oil, Dr. Bronner's Sri Lankan subsidiary, Serendipol, began operation as the world's first major fair trade-certified producer of coconut oil. Serendipol now produces over 1300 metric tonnes of coconut oil for Dr. Bronner's and 300 metric tonnes for other companies. It employs close to 200 staff, including administrative staff and field officers, and buys from more than 400 farmers. Serendipol supports its growers by educating them on organic agricultural methods and by supplying compost to improve the productivity of their land. Compensation and working conditions at Serendipol are far superior to comparable operations in the area. Dr. Bronner's contributions to the project's fair trade fund – reaching over USD 0.25 million in 2010 – are used for a range of projects in health care, education, and staff welfare.[2]

A sister project for palm oil in Ghana (Figure 8.3.2), under the auspices of Dr. Bronner's subsidiary, Serendipalm, has taken longer to emerge. Serendipalm helped about 250 smallholders in Ghana's Eastern Region convert to organic farming, and the palm fruits are processed in a small oil mill that now employs 100 workers. The

---

[1]A video portrait of the project can be seen on http://www.drbronner.com/olive_oil_from_the_holyland.html.

[2]This 2009 video provides a good overview of the project; it can be viewed at http://www.youtube.com/watch?v=-A45lj4ydAs.

Figure 8.3.2 Workers on palm fruit farm in eastern Ghana carrying harvested fruit to the collection point for transport to Serendipalm, the palm oil mill set up by Dr. Bronner's in Asuom, Ghana.
Source: Photo by Alix Audibert, © Dr. Bronner's Magic Soaps.

output currently meets Dr. Bronner's demand; however, since fair and sustainable palm oil for use in natural foods is in high demand, the project is expected to grow in the coming years, providing jobs and attractive returns to farmers in an area left behind by development.

In partnership with Earth Oil India, Dr. Bronner's has helped set up a mint project in Uttar Pradesh, India, to supply the company's mint oil needs and to meet the demand for fair trade menthol by other companies. Finally, Dr. Bronner's purchases domestic fair trade hemp oil from the Farmer Direct Co-operative in Canada, as well as Fair for Life-certified avocado oil from Kenya.

There are other fair trade projects on the company's horizon. A project is planned for the collection of wild jojoba seeds by Seri Indians in the Sonoran desert. The Seri are not farmers, but IMOs Fair Wild program offers fair trade certification of such non-farming projects, ensuring that wild collection is also done in an ecologically sustainable manner.

For each of Dr. Bronner's projects, a "fair trade premium" of 10% is paid on fruits and labor into a fund that is democratically administered by the farmers and workers with input from the company; however, the company does not control the vote. Each fund supports the needs of the community by financing: health clinics, sanitation projects, digging wells, school supplies for children, equipment for hospitals, and whatever else the committee decides is important for the needs of the community. The idea behind the fair trade premium is that companies in the developed world realize the advantages of trade in the form of lower labor costs and taxes in setting up operations in communities in the developing world. However, the same communities often have little or no governmental services and do not have access to clean water, decent schooling, and so on. The fair trade fund helps address these important community needs.

8.3.4 CORPORATE ACTIVISM

Another major focus of Dr. Bronner's is fighting for organic integrity in the US marketplace. Organic agriculture is about cultivating crops naturally and sustainably, utilizing agro-ecological methods to improve yields and address pest and weed pressure, rather than relying on unsustainable petro-chemical-based inputs in the form of synthetic fertilizers, pesticides, and herbicides which pollute food, water, and ultimately human bodies. Unfortunately, the mainstream cosmetics industry is largely built on marketing hype, promising eternal youth and beauty in the form of one magic ingredient or another. Dr. Bronner's believes the intersection of the cosmetics industry and the organic movement has been largely false; many companies are adding token amounts of organic herbs and aloe in formulations that are otherwise based on conventional agricultural materials or petro-chemicals.

Working with the Organic Consumers Association's "Coming Clean" campaign and other authentic organic brands, Dr. Bronner's has worked in the natural products market to get these "Organic Cheaters" to either live up to, or drop, their organic claims. The methods have involved consumer education, public relations, and ultimately litigation. In 2010, Whole Foods, the largest natural products retailer in the USA, decided to address the rampant fraud by introducing guidelines for organic personal care products (see Chapter 12). The move has helped clean up the fraud; in general, consumers can now trust organic claims on personal care products sold in the natural food and body care channel in the USA.

Dr. Bronner's has also long fought for the recommercialization of non-drug fiber and seed cultivars of cannabis, termed "industrial hemp", in the USA. Hemp seed oil is used in all of the company's liquid and bar soaps, where the high content of the omega 3 Essential Fatty Acid in hemp oil provides smoothness and less drying after-feel to the lather. Industrial hemp is widely cultivated in Asia, Europe, and Canada, yet US policy continues to conflate industrial hemp with marijuana. Cultivation and processing is therefore prohibited, although ironically hemp seed and fiber are allowed to be imported into the USA. The contributor (David Bronner) was arrested in front of the White House in June 2012 for protesting against the US ban on industrial hemp cultivation by American farmers. Mr. Bronner placed himself for three hours in a steel cage where he harvested live hemp plants and pressed hemp seed oil under a large picture of President Obama superimposed on a hemp field. In the cage, Mr. Bronner lectured via microphone, stating in part: "Mr. President, Canadian, Chinese and European hemp farmers are laughing all the way to the bank as our government's regressive policy continues to hand them the world's largest consumer market for hemp seed and fiber products; please let American farmers grow hemp."

Dr. Bronner's supports reform of drug policy across the board, to one based on reason and compassion. The company is optimistic that there will soon be a shift, although there has been some disappointment with the current government's administration.

8.3.5 SUMMARY

To conclude, Dr. Bronner's has social enterprise in its corporate DNA. The owners see their company as an activist engine to help promote a better world, and within their sphere of economic influence to do right by their workers and farmers throughout their supply chains. Their social mission is inextricably tied to their business. They hope more and more businesses adopt this model, and are gratified to see that increasingly this is happening.

9

Green Formulations and Ingredients

Judi Beerling

9.1 INTRODUCTION

Television programmes, newspaper articles and Internet scare stories abound regarding the safety and lack of desirability of having so-called 'chemical ingredients' in cosmetics and to some extent in many other areas of life. Thus, a growing number of consumers perceive that 'chemicals' are bad and that 'natural' is inherently better and safer. These consumers actively seek what appear – from the label and marketing – to be more natural or organic cosmetic products. However, there are no laws regulating what can or cannot be sold as a 'natural' or 'organic' personal care product. Some products, in fact, contain only relatively low levels of natural or organic ingredients. Consumers thus have the option to become ingredient experts, researching everything on the Internet before purchasing products, or to rely on third-party recommendations, trusted brands/companies or certification bodies to provide reassurance.

The natural cosmetic industry has woken up to the fact that the future lies in producing genuinely natural or organic products. This is causing headaches for the development chemist who needs to replace high performing synthetics with more natural alternatives. Certification of raw materials, either as natural or organic, is

Sustainability: How the Cosmetics Industry is Greening up, First Edition. Edited by Amarjit Sahota.

one option followed by many ingredient suppliers to make their products more appealing to such formulators.

This chapter serves to highlight the synthetic ingredients generally felt to be undesirable in a natural or organic product. It also covers how truly natural some current products are versus what their marketing claims would lead consumers to believe.

Huge progress has been made in recent years in the number and types of natural raw materials available to formulators. This has become a key area for many suppliers, highlighted by the large number of ingredient launches at major trade exhibitions each year. Sustainability, biodiversity, natural, organic, green chemistry, biodegradability, phytochemicals and antioxidants are just some of the buzzwords that are in vogue. This movement is not a fad or short-lived trend. Suppliers, manufacturers and formulators alike realise the need to reduce reliance on petrochemicals, as well as satisfying the increasing needs of the 'green' consumer.

Before we can move further however, it is important to state some generally accepted definitions of synthetic, natural, naturally derived, nature identical and organic.

9.2 DEFINITIONS

9.2.1 Synthetic Ingredient

A chemically produced substance or material: a substance or material produced by chemical processes and not occurring naturally.

9.2.2 Natural Ingredient

No uniformly accepted or legal definition exists in the cosmetic arena and commercial organisations often produce their own definition. However, one definition that has found fairly wide acceptance [1] is:

> A natural cosmetic ingredient is any material that has been harvested, mined or collected, and which subsequently may have been processed, without chemical reaction, to yield a chemical or chemicals that are identifiable in the original source material.
>
> For the purpose of this statement, 'without chemical reaction' would permit physical processes such as washing, decolourising, distilling, grinding/milling, separation and/or concentration of the material.

To this, it is commonly accepted to add production of cosmetic raw materials by biotechnology/fermentation.

9.2.3 Naturally Derived Ingredient

A naturally derived ingredient is usually accepted to mean one where a natural raw material is used as the starting point of a chemical process to produce a new chemical or chemicals that in themselves may not be available in nature or in the starting material.

9.2.4 Nature Identical Ingredient

This is defined as a substance that has been produced synthetically, not usually from a natural starting material, in order to produce a material that is identical to that naturally occurring in nature.

Examples of common synthetically produced nature identical ingredients are:

- Vitamin E (tocopherol) – naturally occurs in wheatgerm, sunflower and soya.
- Farnesol – naturally occurs in many essential oils such as palmarosa.
- L-menthol – often produced synthetically rather than being extracted from mint oils.
- Ceramides – synthetically produced nature identical ceramide-like structures are better than using animal sources.
- Parabens can also be considered to be nature identical since very similar materials are found in nature.

9.2.5 Organic

In this context, we are talking about natural ingredients that result from organic agricultural methods.

Organic farming is a form of agriculture which avoids the use of synthetic fertilisers, pesticides, plant growth regulators and livestock feed additives. As far as possible, organic farmers should practice crop rotation, integrated pest management, crop residues, animal manures and mechanical cultivation to maintain soil productivity and tilth to supply plant nutrients and to control weeds, insects and other pests. Organic agriculture prohibits the use of Genetically Modified Organisms (GMOs), radiation and similar substances. Organic livestock standards involve a high degree of animal welfare and prevent the routine use of antibiotics and growth promoters.

Organic farming and organic food production is regulated in North America, Europe and other countries like Japan. Growers and producers must meet formal regulations for organic food production. Products made according to these organic regulations are inspected, certified, and allowed to be marketed as 'organic foods'. The term organic is thus legally protected in these countries.

Where regulations exist, it is illegal for a non-certified farm to call itself or its products organic. Farms need to have certification for their crops to be used as 'certified organic' cosmetic ingredients.

9.3 HOW NATURAL ARE CURRENT MARKET PRODUCTS?

High consumer demand for natural and organic cosmetics is leading to a large number of cosmetic products making 'natural' and 'organic' claims, however, how true are these claims? How does a consumer know which products are really natural cosmetics and how accurate are many of the organic marketing claims made today? Organic Monitor [2] undertook research on the ingredient lists of cosmetic brands that had products marketed on their natural/organic features, as well as those having some natural and/or organic component; the research drew some interesting conclusions. Cosmetic products can be rated according to a defined set of criteria, based on those ingredients commonly accepted or rejected by recognised standards for natural and organic cosmetics. By comparing a range of products, Organic Monitor gave brands an overall 'naturalness' score.

A major finding was that most brands studied did not live up to their marketing claims, with some even having conventional cosmetic formulations. Many brands, especially those outside Europe, use a number of synthetic ingredients that are not commonly accepted in natural or organic cosmetics. They were thus classified as semi-natural or 'naturally inspired' although consumers may perceive them as pure natural.

The most authentic products were those certified by a recognised agency. Natural and organic cosmetic standards put pressure on brand owners to use green formulations, otherwise their products cannot get certification. Thus, certified products receive high 'naturalness' ratings. Whilst many non-certified products claim to be natural, the study found few to live up these claims.

The following tables illustrate the ingredient labelling of two skin creams produced by a well-known multinational cosmetic manufacturer. Table 9.1 compares the INCI listing for one of their extensive range of 'conventional' body creams with one of a recently launched 'Naturals' brand that claims to contain at least 95% naturally derived ingredients.

The synthetic ingredients shown in italics, as we will see in the next section, are materials that would be classified undesirable in natural cosmetics. So how does the 'Naturals' product classify itself as '95% naturally derived', when there are ingredients present in italics? These are sodium carbomer (a petro-chemical-based thickening polymer), methylisothiazolinone and phenoxyethanol (synthetic preservatives); it also appears, from the allergen listing and brand marketing, that

Table 9.1 Ingredient lists of two mass market body creams.

Conventional Body Cream	'Naturals' Body Cream
Aqua	Aqua
Glycerin	Glycerin
Paraffinum liquidum	Alcohol denat.
Myristyl alcohol	Cetearyl alcohol
Butylene glycol	Isopropyl palmitate
Alcohol denat.	Glyceryl stearate citrate
Stearic acid	Octyldodecanol
Myristyl myristate	Argania spinosa kernel oil (organic)
Cera Mmcrocristallina	Glyceryl glucoside
Glyceryl stearate	*Sodium carbomer*
Hydrogenated coco-glycerides	*Methylisothiazolinone*
Simmondsia chinensis oil	*Phenoxyethanol*
Tocopheryl acetate	Linalool, limonene, citronellol, benzyl alcohol, *butylphenyl methylpropional, alpha-isomethyl ionone*, geraniol (fragrance allergens)
Lanolin alcohol	*Parfum*
Polyglyceryl-2 caprate	
Dimethicone	
Sodium carbomer	
Phenoxyethanol	
Linalool, citronellol, *alpha-isomethyl ionone, butylphenyl methylpropional*, limonene, benzyl salicylate, benzyl alcohol (fragrance allergens)	
Parfum	

Source: Reproduced with permission from Organic Monitor © 2013.

most likely a – mainly or partly – synthetic fragrance has also been used. (Note: there are certain ingredients present in both products for which it is impossible to tell whether they are from natural sources simply from the INCI name e.g. the alcohol Denat.) The answer to the question lies in the fact that the 'Naturals' body cream has much lower synthetic content, borne out by the fact that these ingredients are all at the bottom of the ingredient list; ingredients have to be given in decreasing order of quantity (apart from when the ingredient is present at less than 1% when it can be in random order). Thus, since typically sodium carbomer would only be used at much less than 1%, as would the preservatives, it is easy to see that the 'unnatural' components would not add up to more than 5% in total. It is interesting to note that the 'Naturals' product has a lower total number of ingredients, which is often a feature of natural formulations where there is often a credo of keeping unnecessary additives to a minimum. It would, of course, also help to reduce cost and complexity. The 'Naturals' product is marketed on claims that 95% of ingredients are from natural origins, no parabens, no silicones, no colourants and no mineral oils, and it is clear that it meets these claims. It is interesting to note that the sensory characteristics of the product are excellent and it would be difficult to tell it apart from its conventional counterparts.

9.4 SYNTHETIC INGREDIENTS NORMALLY ABSENT FROM NATURAL/ORGANIC COSMETICS [3]

It is quite likely that, as consumers become weary of chemical scare stories, cosmetic companies will start shifting their marketing focus to the positive benefits of plant/natural ingredients in cosmetics. This would appear to be a sensible way forward because of the powerful arguments of sustainability – preservation of biodiversity, investment in social community and fair trade projects – as well as the realisation that petro-chemicals are a finite and increasingly expensive resource.

There is actually little peer-reviewed, non-discredited scientific evidence to back up many of the sensationalist media scare stories. However, there is a growing body of evidence that indicates certain chemicals may have long-term health effects. The gap between what cosmetic ingredients are allowed in the USA and Europe is also widening because of the European Union (EU)'s precautionary approach to consumer safety and the Registration, Evaluation, Authorisation and restriction of CHemicals (REACH) chemicals legislation.

Pressure from consumers, activists and marketing departments has led formulators to remove a number of the significant workhorse ingredients used in natural cosmetic formulations. Table 9.2 shows some of the major synthetic or conventional cosmetic ingredients often excluded from natural or organic cosmetics and brief reasons for their lack of acceptability to formulators and marketers of such products.

It is common to see product packaging heralding the brand's green credentials and better safety profile, by virtue of the absence of key synthetics that have created a media issue. This appears as a 'No' list. So 'No sulfates', 'No PEGs', 'No parabens', 'No mineral oil' and so on are very common statements found on these so called 'Free From' products. Some products even go as far as to talk about being 'Chemical Free', which is blatantly ridiculous, as all natural materials are themselves composed of a collection of chemicals. It has to be one of the strangest marketing ploys to trade on the lack of something in a product, sometimes even when it would not have most likely been used anyway! However, those involved in the practice would claim that consumers are looking for this reassurance. It is also true to say that some regulatory authorities, particularly those involved with advertising, are starting to crack down on some of these claims and demand explanation of statements such as 'preservative free'.

Table 9.2 Examples of commonly excluded ingredients.

Class of Chemical	Function(s) in a Cosmetic	Main Reason for Exclusion
Parabens	Preservative	Reputed (unproven) endocrine disruption and general bad press.
Formaldehyde	Preservative and other specialised functions	Known carcinogen
Quaternium 15	Preservative	Formaldehyde donor
DMDM Hydantoin	Preservative	Formaldehyde donor
Silicones	Many functional properties depending on the actual type	Manmade, concerns about biodegradability and skin occlusive properties
Mineral oils, petrolatum and related hydrocarbons	Emollients and carriers	Petro-chemical origin
Synthetic glycols for example, butylene or propylene glycol	Humectants, preservative enhancers and so on.	Petro-chemical origin
Synthetic polymers for example, carbomers or polyacrylates	Thickening, stabilising and many other functions	Petro-chemical origin and concerns in some cases around monomer impurities
Alkoxylated materials, for example ethoxylated or propoxylated ingredients	Various, for example, as surfactants and emulsifiers	Concern about 1,4-dioxane contamination (carcinogenic)
Sulfonated materials	Surfactants	Harsh chemical process
Synthetic fragrance and flavour materials	Perfume/Flavour	Allergies and other reputed health issues
Synthetic dyes	Colourants	Allergies and other reputed health issues
EDTA (ethylene diamine tetraacetic acid)	Chelating agent for heavy metals	Biodegradability issues
BHT (butylated hydroxy toluene)	Anti-oxidant for ingredients and product stability	Endrocrine disruption
Phthalates	Fragrance diluent and solvent (also in plastics)	Human immune system toxicant
Ethanolamines, for example diethanolamine, triethanolamine	pH control and soap emulsifier production. Also as part of other molecules, for example, surfactants.	Nitrosamine formation
Talc	Powder carrier and sensory agent	Asbestos contamination
Chemical sunscreens, for example, octyl dimethyl PABA, homosalate, octyl methoxycinnamate and so on	UV absorbers	Penetration enhancement. Biochemical or cellular level changes. Cancer concern
Sodium lauryl sulfate (for some products)	Surfactant/emulsifier	Skin irritation
Sodium laureth sulfate	Surfactant	See alkoxylated materials above
Saccharin and other artificial sweeteners	Taste improvement in oral care products	Potential cancer risks

Source: Reproduced with permission from Organic Monitor © 2013.

9.5 AVAILABLE GREEN REPLACEMENTS FOR SYNTHETIC COSMETIC INGREDIENTS

Materials which are necessary to the formulation of a cosmetic product largely fall into one of the categories below, according to their function (although individual ingredients may be multifunctional). Examples of more natural/sustainable ingredients are also provided:

a. Emollients

   Emollients are a large class of mostly oily, raw materials that smooth and soften the skin (i.e. act as moisturisers). Petrochemical derived materials have been used for nearly 100 years; for example, mineral oils and petrolatum (often known as Vaseline®).

   Replacement of synthetic or petro-chemical derived emollients by natural oils or naturally derived esters is relatively easy. A wide range of highly effective emollient oils, butters and waxes exist, many of which can be obtained in certified organic grades (Table 9.3). However, the skin feel, spreading, absorption characteristics and stability of the product can be adversely affected. The skill of the formulation chemist is to select the right natural emollients, emulsify them adequately, if appropriate, and keep them stable. This may involve the addition of natural antioxidants to prevent rancidity occurring.

b. Emulsifiers

   An emulsifier is essential to producing creams or lotions by enabling the oil and water to mix and remain stable over a long period of time.

   Beeswax/borax (sodium borate) is a natural emulsifier system that has been used for centuries, but it only produces water-in-oil (W/O) creams. It is extremely difficult to make stable lotions by this process and creams may tend to separate easily.

   The challenge is to find a completely natural O/W emulsifier system that produces elegant emulsions with good sensory properties, similar to those in conventional products.

   Anionic soap systems are used, based on natural fatty acids, but the sodium or potassium hydroxide neutralisers are produced synthetically. However, sodium stearate and other soaps do not usually have good aesthetics or stability.

   Lecithin (a phospholipid extracted from soybean, egg yolk or sunflower seeds) is a good natural option, but can be challenging and expensive to formulate with. However, lecithin is a food-approved and skin-safe ingredient, produced by a gentle extraction process, which improves the moisture balance of skin and hair. Lecithin derivatives, such as hydrogenated lecithin or lysolecithin, provide options to the formulator and easy to use lecithin

Table 9.3 Natural emollient examples.

INCI Name	Source	Benefit
Caprylic/Capric triglycerides	Esterification of glycerol with caprylic (C:8) and capric (C:10) fatty acids, fractionated from coconut or palm kernel oils.	Inexpensive, fast spreading, non-greasy, light skin feel.
Prunus Amygdalus Dulcis (Sweet Almond) oil	Oil expressed from the dried kernels of the sweet almond tree.	Moisturising, relatively light, non-greasy feel. Good lubricity.
Simmondsia chinensis (Jojoba) seed oil	Liquid wax (non triglyceride) obtained from the seed of the Jojoba plant.	Relatively stable oil with soft, silky skin feel. Similar composition to the wax esters found in skin sebum.
Butyrospermum parkii (Shea) butter	Extracted from the fruit (nuts) of the Shea (Karité) tree.	Excellent skin and hair conditioner said to have anti-aging, UV protective and anti-inflammatory properties.
Cocos nucifera (Coconut) oil	Oil derived from the expression of the dried inner flesh of coconut kernels.	Moisturising and skin lubricating properties. Melts at skin temperature making it useful in massage oils.
Olea europaea (Olive) fruit oil	Obtained from the ripe fruit of the olive tree.	Skin conditioning emollient rich in oleic acid.
Helianthus annuus (Sunflower) seed oil	Expressed from the seeds of the sunflower, consisting primarily of triglycerides of linoleic and oleic acids.	Rich in polyunsaturated fatty acids. Low cost carrier oil.

Source: Reproduced with permission from Organic Monitor © 2013.

emulsifier blends are available that can speed up the manufacturing process (Table 9.4).

There are no natural non-ionic emulsifiers, but there are a host of naturally derived materials which are accepted by standards bodies such as the BDIH, Ecocert Greenlife or the Soil Association. Examples of classes of materials include sucrose esters, glucose esters, sorbitan esters, glyceryl esters, polyglyceryl esters, alkyl glucoside blends, inulin lauryl carbamate, olive oil esters and blends thereof.

Anionic emulsifiers are normally produced by esterification of naturally derived feedstock, such as hydrophilic acids (citric, lactic, glutamic, etc.) and fatty alcohols and in some cases fatty acids and glycerine. Examples include glyceryl stearate citrate, acyl glutamates, acyl lactylates, acyl phosphates.

Table 9.4 Natural emulsifier examples.

INCI Name	Natural Source	Benefit
Carthamus tinctorius (Safflower) oleosomes (and) glycerine	Safflower seeds	Cold processing possible, can replace conventional emulsifiers, low irritation/ moisturising benefit.
Inulin lauryl carbamate	Chicory root	Cold processing, good for thin sprayable milks.
Sucrose laurate	Sugar cane (sucrose) and palm derived fatty acids.	Very mild, non-irritating. Anti-irritant properties towards other materials in the formulation, moisturising.
Cetearyl olivate (and) sorbitan Oolivate	Olive oil, coconut oil, sorbitol	Anti-aging, anti-redness and moisturising active and emulsifier with skin compatible liquid crystal structure.
Sodium stearoyl glutamate	Amino acid	Improves the stability of natural emulsions,
Polyglyceryl-3 Polyricinoleate	Glycerine and fatty acids from castor oil	Preparation of w/o creams containing up to 80% water

Source: Reproduced with permission from Organic Monitor © 2013.

c. Surfactants

The most common synthetic surfactant used in foaming products, to cleanse the skin or hair, is sodium laureth sulfate (INCI name), commonly abbreviated to SLES. In the USA, it is also common to see ammonium laureth sulfate. In some cases, sodium or ammonium lauryl sulfate (SLS/ALS) may be added to increase foaming, and a mild surfactant such as cocoamidopropyl betaine (CAPB) used to reduce skin irritation potential. SLES is an ethoxylated raw material and as such is normally excluded from natural formulations. The non-ethoxylated SLS (and its close relative sodium coco sulfate) is actually allowed by many, but not all, of the standards bodies and, similarly, CAPB is permitted currently by Ecocert Greenlife but not by many other natural certification standards.

Alternatives to the conventional SLES/CAPB systems include natural materials with surfactant properties, including saponins from plants sources, such as Yucca (e.g. *Yucca glauca*) or Quillaia (*Quillaja saponaria*). Saponin-rich extracts are, however, often highly coloured and difficult to formulate into an acceptable foaming product. They are also expensive. A certified organic *Quillaja saponaria* Molina is available, which shows good foam stability in hard water. Alkyl polyglucosides (APGs) are the workhorse of many natural cleansing product formulations. This class of mild, biodegradable surfactants,

Table 9.5 Naturally derived surfactant examples.

INCI Name	Natural Source	Benefit
Sodium coco sulfate	Coconut fatty acids	High foaming. Somewhat less irritating than sodium lauryl sulfate
Disodium cocoyl glutamate	Coconut and amino (glutamic) acid from fermented sugar	Mildness, non-drying with skin and hair conditioning properties
Sodium coco-glucoside tartrate	Coconut, glucose and tartic acid	Excellent mildness
Sodium lauroyl lactylate	Lauric acid and lactic acid	Foam booster/stabiliser
Sodium cocoyl apple amino acids	Coconut, amino acids characteristic of apple juice	Mildness, high foaming
Disodium lauryl sulfosuccinate	Coconut oil	Mildness, high foaming. Produces a very tight, dense lather with a pleasant after feel on the skin.

Source: Reproduced with permission from Organic Monitor © 2013.

which includes coco, lauryl, decyl and caprylyl/capryl glucosides, are produced from fatty alcohols (coconut or palm oil-based) and glucose (from corn or potatoes). They are widely considered to be the most natural of the chemically produced surfactants currently available. 'Greener' surfactant blends have been launched which take the hard work out of combining the correct quantities of different naturally derived surfactants (Table 9.5).

d. Silicones

Silicones are a chemical class that encompasses a wide variety of materials; they are synthetic polymers made from silicon and oxygen, two of the world's most plentiful natural elements. Although 'silicone' is often used as a generic term for nearly all substances that contain a silicon atom, it is more properly described as an entirely synthetic polymer containing a Si–O backbone. To this backbone, organic groups are frequently attached to the silicon atoms via a Si–C bond.

Finding suitable natural alternatives to silicones is difficult, since this group is very diverse; they include fluids, emulsions, anti-foams, resins and elastomers, and have a variety of cosmetic applications. However, if we consider dimethicone replacements – materials that impart a lubricious feel to skin and hair products – then there are a few natural alternatives. These include a marine 'vegetable silicone' (INCI – aqua and *Chondrus crispus* extract), a mushroom-derived chitosan derivative (INCI – chitosan succinamide) and a milk thistle extract (INCI – *Silybum marianum* ethyl ester). These are designed to impart a natural silicone-like feel to water-rich products such as emulsions, imparting a non-occlusive protective film to the skin.

Cyclomethicones are another major silicone class used in cosmetics. They possess volatility, and are necessary for products where eventual evaporation of the material is required. Such silicones are used to provide a light, transitory silky skin or hair feel and act as a carrier fluid in for example, antiperspirants and deodorants. In the last few years, several potential naturally derived replacements have become available. The best known is probably is a blend of coconut alkane and coco-caprylate/caprate, a volatile and low odour alkane sourced from vegetable oils. It is a clear, colourless emollient with volatility properties similar to petroleum-derived isododecane and synthetically-derived cyclomethicones.

e. Hair Conditioners

Replacing quaternary ammonium compounds and other cationic compounds for hair conditioning is particularly difficult. However, a new technology has recently been developed by Inolex Chemical Co. USA with the trade name Emulsense[1]. This involves the selection of a certain natural amino acid that has the potential to be selectively esterified. This esterification is performed with L-isoleucine, an essential amino acid derived from the fermentation of non-GMO rice and Brassica-derived long chain fatty alcohol, followed by neutralisation with a fermentation-derived organic acid. The result is a natural cationic emulsifier, a mimic of petro-chemically derived amidoamines (fatty amine quats). Emulsense has the INCI name Brassicyl isoleucinate esylate (and) Brassica alcohol. The process to produce this natural cationic compound follows green chemistry principles. Reactants are 100% renewable and sustainable. No solvents or petro-chemicals of any kind are used in the manufacturing process. The neutralising acid has a dual function: as a neutraliser and catalyst. This means mild reaction conditions with no need for heavy metal catalysts and reduced energy consumption. Production involves a one-step high-yielding process, with water as the only reaction by-product.

Another natural hair conditioning agent is *Pentaclethra macroloba* seed oil which is derived from the *Pentaclethra macroloba* (Pracaxi) fruit. It is an Ecocert-certified material rich in bio-behenic acid. It is extracted from the Pracaxi seeds; this is a wild plant from the Amazon rainforest whose oil is traditionally used by the local people for hair and body treatment.

Inulin, derived from chicory root, is yet another potential natural substitute for cationic conditioning polymers in rinse-off products.

---

[1]Emulsense™ is produced by the Inolex Chemical Company, USA – http://www.inolex.com/cosmetic_ingredients/sensory_emulsifiers/emulsense (accessed July 19th 2012).

f. Polymers

Synthetic polymers are used for many different purposes in a cosmetic formulation. For example, a whole range of polymers, such as the INCI designated 'carbomers' are used for thickening and emulsion stabilisation or suspension of other materials such as scrub or visual effect particles. Different types of polymeric emulsifiers are also relatively common, as are silicone-based polymers, for imparting skin or hair conditioning and other sensory effects. Biodegradability and sustainability are the areas of concern for these materials. A number of natural polymers do exist that can be used as replacements, although they may not be able to replicate the full functionality of synthetic polymers. Examples of materials for thickening and stabilising formulations, are as follows:

- Xanthan gum (fermentation derived)
- *Crondus crispus* (Carrageenan)
- Starch based materials, for example, *Zea mays* (Maize) Starch or tapioca starch
- Microcrystalline cellulose and cellulose gum
- Tara gum (*Caesalpinia spinosa* gum) and acacia senegal
- Sclerotium gum (fermentation derived)
- Succinoglycan gum (fermentation derived)
- Algin (from the seaweed kelp)
- Hectorite clays.

It is often the case that blends of these natural thickeners can be made to produce better effects than a single ingredient alone. The major disadvantage of many of these materials is the potential for a sticky skin feel or stringy appearance to the product. Achieving clarity in surfactant blends and gels is another challenge. There are a number of new or improved materials coming into the market to overcome some of these disadvantages.

g. Solvents

Propylene glycol is a petroleum derivative that is believed to penetrate the skin and possibly weaken protein and cellular structures. Glycols, such as propylene or butylene glycol, are commonly used to extract botanical materials. Many companies now offer the same extracts in other suitable solvents such as glycerin, water, alcohol or in a natural oil base.

As a renewable 'green' replacement for glycols, a material with the INCI name Propanediol, is produced from corn sugar by fermentation. It has also been shown by two independent studies to lack irritation and sensitisation potential, even at high concentration. Biofermentation-derived butylene and pentylene glycols are now also available.

h. Actives

A large number of prestigious skin and hair care products rely on high-tech synthetic actives, such as peptides and vitamins, for their efficacy and

Table 9.6 Examples of sustainable natural active ingredients.

INCI Name	Source	Benefit
Oat beta glucan	Oat (*avena sativa*) kernels	Hyaluronic acid replacement with moisturising and skin rejuvenating properties
Hydrolyzed avocado protein and maltodextrin	Defatted avocado pulp	Anti-aging, plumping and detoxifying
Bidens pilosa extract, *Elaieis guineensis* (Palm) oil, *Gossypium herbaceum* (Cotton) seed oil, *Linum usitatissimum* (Linseed) seed oil	*Bidens pilosa* (Hairy Beggarticks) plant and essential fatty acid rich vegetable oils	Anti-aging, skin lightening, anti-acne, retinoid-like activity
Glycerin, *Leontopodium alpinum* meristem cell culture, xanthan gum	Biotechnological plant stem cell culture technique using the Edelweiss plant	Reduction in wrinkle depth. Anti-oxidant activity
Butylene glycol, *Enantia chlorantha* bark extract, oleanolic acid	*Enantia chlorantha* bark extract (from an African tree) rich in protoberberines and oleanolic acid	Reduces pore size, mattifying and refines skin texture. Regulates sebum production
Bellis perennis (Daisy) flower extract	Daisy flowers	Skin lightening/brightening, reduction in age spots

Source: Reproduced with permission from Organic Monitor © 2013.

product claims. Fortunately, an ever increasing number of naturally based active ingredients whose performance has been substantiated by *in vitro* and *in vivo* trials are now available to the formulator. In the case of vitamins, naturally derived versions are often available or they can be found in selected natural extracts. Table 9.6 shows some examples of anti-aging natural active ingredients, some of which are produced by biotechnology or new techniques that may not even involve agricultural production. However, this list could run to many pages since actives are an area where there is a huge amount of choice to the formulator.

i. Antioxidants

Alternatives to synthetic antioxidants for product protection include natural Vitamin E components (alpha-tocopherol) and vitamin E rich oils. Rosemary extract or oil is also commonly used. For antioxidant benefits in the product, there are a huge number of natural ingredients available. These include natural extracts such as avenanthramides from oats; pomegranate polyphenols (ellagic acid and punicalagin); malted barley extract rich in the polyphenol derivatives oligomeric proantho-cyanidolins (OPC) and ferulic acid; and *Sesamum indicum* (sesame) oil unsaponifiables, rich in sesamine and sesamoline.

j. Preservatives

Preservation is probably the major issue that companies formulating natural, and in particular organic products, have to wrestle with. The situation is complicated by the differing laws governing their use in the EU, the USA and other countries. The situation is further exacerbated by the different preservatives (or antimicrobials) that are certified by the different international standards agencies (see Chapter 10).

Options include the synthetically produced 'nature identical' preservatives, plus some 'natural preservatives' that are accepted by most of the certification bodies. Examples of the natural identical types include benzoic acid/sodium benzoate, sorbic acid/potassium sorbate, salicylic acid and benzyl alcohol. These occur naturally in plants and are on the EU permitted list; they maybe used in natural cosmetics, even though they are synthetically produced.

Glucose (and) lactoperoxidase (and) glucose oxidase (INCI name) is a two-part enzyme system, which mimics a preservative technique used in nature. The system is permitted by a number of certification standards but uncertain GMO status of the glucose has led to the removal of the commercial material by the Soil Association.

There are a number of natural blends, fragrance ingredients and plant extracts that can be used to boost preservative efficacy or to replace conventional preservatives in certain formulations (but they cannot be claimed as such). Examples include *Citrus grandis* (grapefruit) fruit seed extract, Japanese Honeysuckle extracts, spice extracts and numerous essential oil blends. Some of these materials have, however, been a source of controversy due to their less than natural production method, adulteration with synthetic preservatives (grapefruit seed extract) or the presence of natural paraben structures (honeysuckle extract).

Denatured alcohol is used by some companies as a preservative, since 15–20% in aqueous based products makes them self-preserving. In combination with other self-preservation techniques, it is possible to reduce this level, making products more consumer-acceptable. However, some consumers who are concerned about the skin drying effects of alcohol may shy away from products that list this material on the label.

One strategy that companies have used to avoid including synthetic preservatives is to use alternative 'multi-functional' materials. These are not listed in the EU Cosmetic Regulation (Annex V) of approved preservatives. They are added for another primary purpose, for example moisturisation, but have the secondary benefit of reducing the possibility of microbial spoilage. However, the legality of this approach has been questioned, particularly where products claim to be 'preservative-free'. Examples of popular materials that are used in this manner are: glyceryl caprylate, glyceryl caprate, glyceryl undecylenate, phenethyl alcohol and sodium anisate/sodium levulinate.

Preservation a highly complex issue and strategies will need to be based on a number of factors, such as the type of product formulations being produced, manufacturing methods, cost and whether certification is desired in the finished product.

k. Fragrances

The main alternative to using synthetic fragrances is to incorporate blends of essential oils, oleoresins, absolutes and other natural aromatic compounds that are used in the fragrance industry. These blends will still be counted as fragrances and should conform to International Fragrance Association (IFRA) regulations. Fragrance allergens will also need to be identified and labelled in the same way. To reduce cost, blends of essential oils and natural identical synthetic aroma chemicals can be produced. This broadens the scope of stable fragrance types that can be created, but would not be considered natural for certification purposes.

Absolutes, concretes and resinoids are highly concentrated fragrance ingredients that add strength and depth to some natural perfume types. Those extracted with anything other than ethanol or supercritical carbon dioxide are currently allowed by certain private standards, such as COSMOS. This is because of the difficulties in obtaining these materials in more natural forms. For COSMOS, these ingredients may not be certified as organic and thus are only for use in cosmetic products with natural certification. Solvents (often petro-chemicals such as hexane) are required to be recycled and removed from the extract to the highest degree technologically (or technically?) possible.

l. Colours

The easy route is to exclude all dyestuffs, but products can end up looking bland or off-white. Often the consumer will associate 'E numbers' on foods with synthetics but in fact, natural dyes often also have E numbers!

A number of oil or water soluble natural colours, usually coming from the food industry, are available to the cosmetic formulator. Natural colours are derived mainly from plants, for example seeds (e.g. annatto); roots or rhizomes (e.g. turmeric); sugars (caramel); leaves and stems (e.g. spinach); flowers (e.g. lutein from Tagetes); vegetables (such as red cabbage, beetroot juice and carrot oil extract); fruits (grape juice); algae (beta carotene). The red colour carmine is also technically natural but comes from an animal source – the cochineal beetle – so it tends to not be acceptable for this reason. It has also been linked with some allergic reactions.

Work has also been carried out by suppliers of natural colours to ascertain the optimum pH range and other aspects of the formulation to ensure improved light and temperature stability in products. Some of these materials have colour index numbers and so can legitimately be used to colour cosmetic products.

For decorative cosmetics, the colour for lipsticks, eye-shadows and similar products come from the use of water (and usually oil) insoluble colourants, such as iron oxides, lakes and/or pearlescent pigments. A number of Ecocert Greenlife approved mineral pigments, interference pearls and mica-based glitter particles now exist. This list is growing continuously and will make the production of more natural, but still colourful, decorative cosmetics much easier to achieve.

m. UV Absorbers/Sunscreens

A fierce debate is currently raging on the safety of the so-called 'organic' or 'chemical' sunscreens. A number of these are reputed to be penetration enhancers and some, including oxybenzone, are thought to be endocrine disruptors.

There are, as yet, no natural, plant derived sunscreen agents allowed by EU or US legislation. The EU has a positive list of UV absorbers that can be used in products making a SPF or other sunscreen claim. In the USA, the situation is even more complex as sunscreens are regulated as over-the-counter (OTC) drugs and the Final Monograph contains even less UV absorbers than the EU list. Getting a new chemical sunscreen, let alone a natural UV absorber, tested and accepted on either, or both, of these lists is prohibitively expensive and would most likely require extensive animal testing.

Companies can make use of plant extracts, such as black tea, Peruvian maca (*Lepidium meyenii*), green coffee, shea butter and other plant extracts rich in UV absorbing phyto-chemicals. These can boost SPF efficacy but they cannot be used as the sole sunscreen ingredient. Most manufacturers of 'natural' sunscreen products utilise inorganic UV protective materials, such as titanium dioxide (TiO_2) and zinc oxide (ZnO). These are produced from mined ores and can therefore loosely fall under most definitions of natural. However, to make an aesthetically pleasing effective product it is much easier to use them as dispersions in a suitable emollient oil. For natural cosmetics, these emollients need to be acceptable natural oils or naturally derived esters without any synthetic additives. The market currently only has a few such products but the range is steadily growing.

n. pH adjustment

Sodium or potassium hydroxide can be used to raise the pH in natural formulations, but will be counted as synthetic for certification purposes. A more natural, albeit more expensive, alternative is to use the amino acid L-Arginine. This material has a pH of around 10.5 and also imparts skin moisturising properties to a formulation.

Alpha hydroxyl acids, such as lactic or citric acid, from natural sources can be used to adjust the pH downwards, that is to acidify the product.

o. Chelating Agents.

Ethylene diamine tetraacetic acid (EDTA) has been the material of choice for many years. However, it is not natural and has environmental biodegradability and suspected health issues. The main replacement for EDTA as a chelating agent is phytic acid (found in the hulls of nuts, seeds and grains) or its sodium salt, sodium phytate. Trials show an excellent chelating effect, in some cases superior to EDTA, but the cost is considerably higher.

Another possibility is ethylenediaminedisuccinic acid (EDDS). Although not natural, EDDS is made from aspartic acid and has excellent biodegradability, unlike EDTA. The trisodium salt of EDDS is commercially available[2] and is approved by the EU Ecoflower and Nordic Swan for detergents and household cleaners. Some companies have chosen to simply use aspartic acid or its sodium salt for this purpose as it is naturally derived.

9.6 FORMULATION ISSUES WITH GREEN INGREDIENTS

The performance of functional natural materials, for example preservatives, may not (yet) match that of conventional ingredients. Hence, the formulator may need to consider blending ingredients or accept that there will be some impact on performance in certain areas. There is unfortunately no 'one size fits all' answer for example, to preserving natural or organic products, which are by their nature made from materials that bacteria find highly nutritious. In other areas, such as emulsification and actives, there are now a number of highly effective natural ingredients on offer. Once the move is made to sourcing of organic materials, however, there is much less choice in high-performing ingredients, something which committed consumers of organic products have come to accept.

Problems can also occur when scaling up from laboratory to manufacturing, and even from batch to batch. Close quality control checks need to be in place to ensure that all batches of the natural raw materials and the finished product meet the required specifications (colour, pH, odour, etc.) and pass microbial challenge tests on a regular basis.

Some materials can also be more difficult to handle or work with. For example, alternative preservatives or certain essential oils or botanical materials may need to be refrigerated until use.

9.7 SUMMARY

The natural cosmetic industry has recognised that the future lies in producing genuinely natural and/or organic products. This is creating major challenges for the development chemist who still needs to formulate high-performing products with more natural alternatives. However, these same formulators now have an increasing palette of raw materials to work with. This has become a key area for many suppliers

[2]Natrlquest E30 is produced by Innospec Corp. USA.

that realise this is not just the current 'fashion'. Virtually all areas of the cosmetic industry recognise the need to reduce reliance on petro-chemicals, as well as the need to meet the demands of the growing band of 'green' consumers. Certification of raw materials, either as natural or organic, is one important mechanism used by many raw material suppliers to improve their natural or organic credentials.

However, over-riding all of the marketing hype surrounding natural/organic cosmetics and their ingredients is the true desire of many companies to operate in a more sustainable manner. Sustainable raw materials, produced from renewable feedstock, ideally from a waste by-product of another industry (such as food or textiles) and only modified by 'green chemistry', low energy processes, or produced by biotechnology, are leading the way in this 'brave new cosmetic world'.

REFERENCES

[1] Warby, J.R. (Jan. 2011) Australian Society of Cosmetic Chemists, Position Paper, http://www.ascc.com.au/papers.php?id=7 (accessed July 19 2012).
[2] Organic Monitor Ltd. (2011) Technical Insights Report on Natural & Organic Cosmetics Brand Assessment, London, UK.
[3] Organic Monitor Ltd. (2009) Technical Insights Report on Natural & Organic Cosmetic Ingredients, Formulation Challenges & Solutions, London, UK.

10

Green Standards, Certification and Indices

Judi Beerling and Amarjit Sahota

10.1 INTRODUCTION

One of the most debated issues in the cosmetics industry is the role of standards and eco-labels. Although the number of standards and eco-labels is increasing, there is fragmentation with few schemes gaining popularity outside their home countries. Eco-labels attempt to communicate certain aspects of sustainability to consumers. They are typically voluntary schemes awarded by third-party organisations to products and services that meet a certain level of environmental performance.

Eco-labels related to natural and organic cosmetic standards have the highest take-up rate in the cosmetics industry, however these standards are based on green formulations. No standards and eco-labels exist specifically for the cosmetics industry that cover wider aspects of sustainability, such as packaging, carbon and water footprints, resource usage, waste materials and so on.

In the area of natural and organic cosmetics, the industry is calling for regional – if not international – standards; however only a few standards have developed a trans-national presence. The major natural and organic standards in Europe, North America and other regions will be covered in this chapter, as well as regional 'harmonisation' initiatives such as COSMOS and Natrue. The use of fair trade labelling schemes in the cosmetics industry will be discussed, as well as other

Sustainability: How the Cosmetics Industry is Greening up, First Edition. Edited by Amarjit Sahota.

eco-labels and sustainability indices. Eco-labels are becoming ubiquitous in the cosmetics industry; almost all countries or regions have some labels, standards or indices that represent some form of environmentally-friendly products.

10.2 NATURAL AND ORGANIC COSMETIC STANDARDS

Consumers face a major problem when selecting natural and organic cosmetic products. They look at product labels, but there are no laws regulating what can or cannot be marketed as a 'natural' or 'organic' cosmetic product. Indeed, many products are often marketed as such but contain relatively low levels of natural/organic ingredients. Thus, standards are becoming important with certification providing some assurance to consumers that products meet certain a level of 'naturalness' or organic content.

There are no national or regional regulations for natural and organic cosmetics, as there are for organic foods. A number of cosmetic standards look to organic food regulations as their starting point. However, organic food ingredient production techniques do not cover many areas of chemistry required for cosmetic ingredients. Thus, each certification agency has its own interpretation of what constitutes an organic or natural cosmetic product, and permissible ingredients. As a consequence, some fundamental differences exist between these private standards.

The proliferation in number of natural and organic cosmetic standards is creating problems for the development chemist who needs to formulate natural and organic cosmetic products, often with certified ingredients. Certification of the finished product can be a time-consuming and often expensive process. Certification of raw materials, either as natural or organic, is one option open to many ingredient suppliers. In some cases, the process is undertaken by the finished product manufacturer which relies on the raw material supplier to provide adequate, often sensitive, information to the certification agency.

There are three major standards – USDA/NOP, NSF ANSI 305 and NPA in North America. However, the adoption rates remain low; Organic Monitor estimates that less than 10% of natural and organic cosmetics are certified in the region in 2011 [1]. The Natural Products Association (NPA) standard is the frontrunner, adopted by some of the leading natural cosmetic firms in the USA. In February 2009, the NSF ANSI 305 'contains organic' standard was announced. However, relatively few products were certified according to this standard in 2012. The absence of a 95–100% organic cosmetic standard is leading many companies to adopt the USDA/NOP standard. Although designed for organic food products, it is popular with companies that make relatively simple, organic cosmetics from food-grade ingredients.

Natural and organic cosmetic standards are also being introduced in other regions, such as Australasia and Latin America. The private standards in these regions have exceptionally low adoption rates, partly because natural cosmetic firms have an export focus and prefer to adopt European standards. Australia has the distinction of having a national standard for organic cosmetics; however, as for organic foods, it remains a voluntary standard.

Few natural and organic cosmetic standards have developed a regional – let alone international – presence. There was much anticipation of the COSMOS initiative, developed by the leading standards bodies in Europe. After nine years of negotiations, it was finally implemented in January 2010. The initial aim was to make COSMOS a harmonised pan-European standard. However, it faces growing competition from the Brussels-based Natrue label. The Natrue standard has become popular, partly because it is backed by leading natural cosmetic firms that include Weleda and Lavera.

The slow emergence of regional standards makes a global standard highly unlikely at least for some years yet. Standards are mainly adopted at national levels, and although based on similar philosophy the differences between some leading standards remain significant. Ecocert Greenlife continues to gain an international presence although it is one of the major partners of the COSMOS European harmonisation initiative. The standard is gaining currency across the globe, adopted by companies in the Americas, Asia, Australia, as well as in Europe.

10.2.1 Major European Standards for Natural and Organic Products

Natural and organic cosmetic standards are the most widely adopted in Europe. The region has over 20 private standards for natural and organic cosmetics; the leading ones are Natrue (Belgium), BDIH (Germany), Ecocert Greenlife (France), CosmeBio (France), Soil Association (UK) and ICEA (Italy). Ecocert and BDIH are the leading standards, partly because of the high adoption rates in France and Germany respectively. Research by Organic Monitor shows that over 70% of natural and organic cosmetics are certified in these two countries [1].

10.2.2 BDIH (Germany)

BDIH is the Association of German Industries and Trading Firms for pharmaceuticals, healthcare products, food supplements and personal hygiene products. Located in Mannheim, it has over 500 members of which around 125 are natural cosmetic and personal care product companies. It developed standards for certified natural cosmetics in 1996, introducing them in spring 2001. Leading natural cosmetics producers, such as Weleda, WALA-Heilmittel, Laverana, Logocos and Primavera, worked with BDIH to develop this standard.

BDIH is the leading certifier of natural cosmetics in Europe. Natural cosmetic products made according to its standards are given the 'BDIH Kontrollierte Natur-Kosmetik' logo, which means certified natural cosmetic. Due to the widespread success of German natural cosmetics, the BDIH logo is the most evident logo for natural cosmetics in Europe.

BDIH is not involved in certification of organic products; however, it does verify the organic status of raw materials used in its certified natural products. BDIH established the International Organic and Natural Cosmetics Corporation (IONC GmbH) to organise global checks on cosmetics to ensure that they meet the BDIH Standard. The basic requirements of BDIH standards are available from the IONC web site [2].

The key points of the BDIH natural cosmetics standard are:

- Raw materials should be natural and obtained from plants, preferably organically cultivated or from controlled wild collection. Minerals are also generally allowed. Irradiation and genetic modification are prohibited.
- Only certain 'mild' chemical processes are allowed using prescribed types of natural feedstock.
- Some synthetic preservatives are permitted but must be quoted on the label.

BDIH is one of the key agencies involved in developing the European COSMOS standard.

10.2.3 Ecocert Greenlife (France)

Ecocert Greenlife is a leading certification and inspection agency for organic products in Europe. Established in 1981, it has grown to have offices across the globe. Although it is active in over 80 countries, its major presence is in France where it certifies about 70% of organic products.

Ecocert promotes organic farming practices, sustainable ingredient sourcing and other environmental initiatives. It also encourages these initiatives by its natural cosmetic and organic cosmetic standards. Ecocert is the only certification agency in Europe to have natural and organic cosmetic standards; the key points are:

For both organic and natural cosmetics, there must be 95% minimum of approved natural or naturally derived ingredients. Thus, a maximum of 5% of certain acceptable synthetic products (normally preservatives) are allowed.

For a certified natural cosmetic, there must also be 5% minimum of organic certified ingredients (of the total ingredients), which represents at least 50% of the vegetable (plant) ingredients.

Additionally, for a natural and organic cosmetics, at least 10% of the entire formula must be certified organic unmodified plant materials. In this case, 95% of the organic ingredients present are required to be certified. Water is accepted in the natural content calculation.

Ecocert standards for natural and organic cosmetics are the most widely adopted globally. One of the reasons behind the high adoption rate is that Ecocert organic standards are the most realistic to achieve, especially when compared to other 'stricter' organic standards, such as the Soil Association or USDA/NOP. Its critics argue that it is too easy to achieve. However, another factor is that it is one of the few standards to have both a natural and organic standard. In comparison, many certification bodies have just an organic standard.

Ecocert certifies raw materials as well as finished products. A list of approved chemically modified, natural or organic origin ingredients is available to 'those associated with Ecocert' via their password-protected web site. Ecocert has been very actively involved in the development of the European COSMOS standard.

10.2.4 CosmeBio (France)

CosmeBio is the professional association of ecological and organic cosmetics. It was established in 2002 by leading natural and organic cosmetic manufacturers, Ecocert, and the French Ministry of Agriculture.

The CosmeBio labelling scheme involves giving certified organic and natural personal care products the BIO and ECO labels respectively:

I BIO label – the same basic requirements as for Ecocert Greenlife organic cosmetics.
II ECO label – the same basic requirements as for Ecocert Greenlife natural cosmetics.

Certification is undertaken by certification agencies, such as Ecocert and Qualite France. The advantage of the CosmeBio labelling scheme is that certified organic and natural products are clearly distinguishable. However, the adoption rate of CosmeBio labelling scheme is very low outside France.

10.2.5 Soil Association (UK)

The Soil Association is the leading certification agency for organic products in the UK. Established in 1946, the Soil Association operates as a charity and its subsidiary is involved in certification. It was the first organisation in the world to develop an organic certification system in 1967. Although it mainly certifies organic farmland

and organic foods, the Soil Association also has standards for textiles, forestry and cosmetics.

The Soil Association established organic health and beauty (cosmetic) standards in 2002. The standard only covers organic cosmetic products and is based on its agricultural/food background. It is the most widely adopted standard for organic cosmetics in the UK. The key points of the Soil Association standards are:

- To label a product as 'organic', it must contain at least 95% organic ingredients by weight (of the agricultural ingredients) and only permitted non-organic ingredients, processing aids and water.
- If water is added you cannot obtain 100% organic status (water is counted as neutral).
- Products that contain 70–95% organic ingredients can be labelled with the phrase 'product containing X% organic ingredients'.

The Soil Association is also part of the European COSMOS grouping.

10.2.6 ICEA (Italy)

An organic cosmetics standard was introduced by Instituto per la Certificazione Etica ed Ambientale (ICEA) and the Associazione Italiana per l'Agricoltura Biologica (AIAB, Italian Organic Farming Association) in April 2002. The ICEA standard focuses on the materials that it does not allow, which are given in Annex 1 of its standard. Packaging is also regulated. ICEA is one of the founding members of the European COSMOS group.

The ICEA organic cosmetic standard is the most widely adopted in Italy, but is little known outside its home country.

10.2.7 COSMOS

The initiative to harmonise natural and organic cosmetic standards in Europe began in February 2002 when leading certification agencies met at the BioFach trade show in Nuremberg (Germany). The original six collaborators initially included Bioforum Belgium (which later withdrew), the Soil Association, Ecocert Greenlife, CosmeBio, BDIH and ICEA.

The certification agencies recognised that some form of harmonisation of standards was needed to improve the use of organic ingredients and promote wider acceptance of environmentally friendly products. The different standards were felt to be confusing for consumers, bringing the need for a common definition of an organic and natural cosmetic. A single standard would therefore be beneficial to both consumers and manufacturers. The objectives were to encourage suppliers to adopt green chemistry, greener processes and the use of renewable raw materials, to guarantee organic integrity and provide transparency for consumers. The key issues they addressed were that cosmetics ingredients have different natural origins, and thus needed to be considered separately as:

- Physically processed agro-ingredients.
- Chemically processed agro-ingredients.
- Water.
- Minerals.
- Synthetic ingredients (petro-chemicals).

There were protracted discussions on what physical and chemical processes to allow. Other issues, which appear to have prolonged the harmonisation over eight years, included the definition of a natural cosmetic (which some of the parties had never been involved with previously), and how to calculate the organic content and the percentage of organic ingredients necessary for the product to labelled organic. In addition, common procedures for verification, inspection and rules for labels and claims were heavily debated.

The final standard was introduced in January 2010. However, in order to allow for an orderly change over to COSMOS, the private standards are still continuing alongside the COSMOS standard until the end of 2014. After this date, products already certified to one of the member organisation standards can continue to be marketed, but any reformulation would require the company to then apply the COSMOS standard. A COSMOS 'signature' is used under the chosen member certification logo to show compliance.

Some of the key points of the COSMOS standard are:

- At least 20% of the total product must be organic (i.e. including water), but there is an exception of 10% for rinse-off products, watery lotions (such as skin tonics or body sprays) and powders.
- At least 95% of physically processed agro-ingredients must be organic (and must be organic if available). By 36 months after the standard is introduced, the remaining 5% must be organic if available.
- At least 30% of chemically processed agro-ingredients must be organic by the end of 2014.

Approval of raw materials to the COSMOS standard is underway but at a fairly slow rate. Many ingredient suppliers perhaps feel that they do not yet need to upgrade from their current standard, or perhaps they find it hard to get to grips with the requirements of the new standard.

10.2.8 Natrue (Belgium)

Natrue was established in December 2007 by some of the leading international producers of natural cosmetics. The founder members were Weleda, WALA-Heilmittel, Logocos, Laverana, Santaverde and Primavera. These companies were members of BDIH; they set up an office in Brussels so they could be close to the European parliament for lobbying purposes.

The aim of Natrue is to have a voice in the development of clear regulatory definitions for both natural and organic cosmetics, and subsequently an international system of labelling for organic and natural cosmetics. Natrue changed its legal entity from European Economic Interest Grouping (EEIG) into an AISBL (International Not-for-Profit Organisation) so that it could operate on a more global basis.

Natrue introduced standards for natural and organic cosmetics in September 2008. The definition of 'natural' (and subsequently organic) has been derived by Natrue by having discussions with scientists associated with the IKW (the German Cosmetic, Toiletry, Perfumery and Detergent Association). Companies producing conventional cosmetics participate on the scientific board of Natrue. Natrue has three labels for certified products:

I Natural Cosmetics – No organic content is required. Minimum levels of natural and maximum levels of 'derived natural' content (or chemically modified materials allowed under the standard) are specified by product type.

II Natural cosmetics with organic portion. The product must contain (based on the whole formulation) at least 15% of chemically unmodified natural substances and maximum 15% of 'derived natural' substances. Again, these levels increase for certain product types. Additionally, a minimum 70% of the natural substances of plant and animal origin must come from controlled organic farming and/or from controlled wild collection.

III Organic cosmetics. Depending on product type, these must also contain a minimum 20% of chemically unmodified natural substances and maximum 15% of 'derived natural' substances. In this case, a minimum 95% of the natural substances of plant and animal origin must come from controlled organic farming and/or from controlled wild collection.

Although Natrue has set standards for natural and organic cosmetics, certification is undertaken by approved third-party organisations.

In February 2009, Natrue entered a standards equivalency agreement with Quality Assurance International (QAI), a leading certification agency in the USA. The equivalency agreement means that Natural Cosmetics with Organic Portion products meet requirements of the NSF 'contains organic ingredients' cosmetic standard in the USA, and vice versa.

Natrue has been active in trying to reach similar agreements with other bodies worldwide, but talks with the Natural Products Association (NPA) in the USA failed to come to fruition. Thus, in February 2011, NSF and Natrue announced their intention to develop a new natural cosmetics standard for the USA.

10.2.9 Other European Standards

Many other certification agencies have developed natural and organic cosmetic standards for cosmetics. Most have been developed by agencies involved in certification of organic foods. The adoption of these standards is much lower then the aforementioned, and are again mostly on a national basis.

The Non Food Certification Company (NFCC) is a wholly owned subsidiary of the UK organic certification agency, the **Organic Food Federation.** The organisation introduced organic cosmetic standards in 2003 (which were updated in 2007). The use of functional ingredients not of plant origin is quite restricted, although phenoxyethanol is allowed as a preservative (with justification). The rules of organic production and criteria for wild designation of plant materials are identical to the EEC Regulation for organic agricultural products 2092/91. Some companies in the UK work with this certification agency as well as, or in preference to, the Soil Association.

Nature et Progrès, a leading certification agency for organic agriculture and organic products in France, introduced standards for organic cosmetic products in 1998. Certified products are given the Nature et Progrès label.

Many other organic certification agencies that specialise in organic farmland and foods have introduced standards for cosmetics. These include **Organic Farmers and Growers** in the UK and **Naturland** in Germany. The largest number of such standards is in Italy; they include Consorzio per il Controllo dei Prodotti Biologici (CCPB); COsmesi COntrolla NATurale (Co.Co.Nat); Società di Certificazione (SoCert); Bioagricert; Ecogruppo and Suolo e Salute.

10.2.10 Major North American Standards

Compared to Europe, the landscape for organic and natural standards at first sight appears much less crowded. However, when you factor in that major European certifiers also operate there, such as Ecocert Greenlife, a confusing array of logos are present. There have also been a number of high profile lawsuits in this area involving some of the authentic organic brands and the Organic Consumers Association to try to force some of the brands making organic claims to 'clean up their act'. There have additionally been attempts to put pressure on the government and its relevant agencies to take a tougher stance on organic cosmetic labelling, without much success. The Food and Drug Administration (FDA) is powerless to enforce the United States Department of Agriculture National Organic Program (USDA NOP) rules. The only legislation in this area is in the state of California, which has the 2003 Californian Organic Products Act (COPA). This sets a minimum requirement of 70% organic content for organic cosmetics and personal care products, but the other 30% is not significantly restricted.

10.2.11 USDA/NOP

The USDA introduced the National Organic Program (NOP) in 2000. The NOP was enacted as federal legislation in October 2002. NOP regulation, 7 CFR Part 205, regulates the use of the 'organic' term on agricultural and food products. The regulation gives clear guidelines on what organic products can carry the USDA/NOP organic seal. The NOP has four categories of organic agricultural products:

- 100% organic: Only organically produced ingredients are in the product.
- Organic: Contains 95 to 99% organically produced ingredients.
- Made with organic ingredients: At least 70% organic ingredients.
- Some organic ingredients: Below 70%.

Only the first two types of organic products – 100% Organic and Organic – can use the coveted USDA/NOP organic seal. Although the use of the 'organic' term is regulated, the use of the organic seal is voluntary.

Although the NOP has been designed for organic agricultural food products, the labelling scheme is being used for non-food products that contain organic ingredients. The USDA allows cosmetics and other non-food products to bear the USDA/NOP organic seal, provided they comply with its organic standards. In other words, organic non-food products can become certified if they meet the same criteria as organic food products.

The USDA/NOP is generally perceived as the 'gold standard' for organic products in the USA. However, like the Soil Association in the UK, the background of the USDA is in agriculture and food products, thus the USDA/NOP standard is highly restrictive for cosmetics. Many cosmetics companies are therefore adopting other standards designed for cosmetic products like Ecocert Greenlife. In such cases, the USDA/NOP seal cannot also be used even though the products contain certified organic ingredients.

10.2.12 NSF International

NSF was originally founded as the National Sanitation Foundation in 1944. Its name was later abbreviated to NSF International; the independent, not-for-profit organisation is a leading international provider of public health and safety risk management solutions.

After several years of development and a public consultation, NSF International launched its organic personal care standard in February 2009. The NSF/ANSI 305 standard was developed in accordance with American National Standards Institute standards (ANSI). ANSI is a private organisation that administers the US voluntary standardisation assessment system.

The NSF/ANSI 305 standard allows a 'Contains Organic Ingredients' designation for cosmetic products with organic content of 70% or more. It allows products to be labelled 'Made with Organic (up to three specified) Ingredients' or 'xx% Organic'. Various processes and associated non-organic reagents/catalysts, as well as certain synthetic preservatives, are permitted that are not allowed under the USDA/NOP 95% or 100% organic seal. The main features of the standard are:

- Organic forms of chemically processed ingredients must be used if commercially available.
- A minimum 70% organic content by non-water/non-salt weight is required.

This standard is in many ways quite similar to the UK Soil Association organic cosmetics standard.

10.2.13 NPA (Natural Products Association)

Established in 1936, the Natural Products Association (NPA) is the largest and oldest non-profit organisation representing the natural products industry in the USA. It comprises over 10 000 manufacturers, distributors and retailers of natural products.

The NPA launched its natural cosmetics standard in May 2008. It is easy to work with since it publishes a list of approved ingredients by INCI name that products may contain 'by way of example'. Thus, addition of new materials is possible. NPA also certifies raw materials as natural, publishing this information on its web site.

The main aspects of the standard are:

> ....all personal care products labelled or branded 'natural' must be made with at least 95% all-natural ingredients and contain only synthetic ingredients specifically allowed under this standard that do not have suspected human health risks.

It allows for ecological, that is, minimal, processing of natural source ingredients, sometimes known as 'kitchen chemistry'. There is a temporary allowance for certain types of ingredients, such as certain preservatives and quaternised hair conditioning ingredients, which would otherwise not be allowed under the standard. A cynic may say that this may be due to the fact that some of their largest licensees tend to use these materials in their products.

10.2.14 Standards in Other Regions

10.2.14.1 Latin America

Instituto Biodinâmico (IBD) is one of the leading certification agencies for organic agriculture and organic foods in Latin America. It introduced its standard for natural and organic cosmetics in 2007. The Body Care Certification Program of the Brazilian agency is the most widely adopted in Latin America. The IBD cosmetics standard has three levels: organic, made with organic ingredients, and natural. Certified products carry three different IBD seals, which are easily distinguishable.

There are a number of similarities between the IBD and Ecocert standards. However, the biggest difference is the interpretation of composition in order to classify it as organic or natural. IBD does not consider water as an ingredient; organic products are either 'made with organic ingredients' at 70%, or 'organic' with 95% or more organic ingredients (the same calculation system as the Soil Association is used). Thus, a number of products considered organic by Ecocert would be considered natural by IBD.

In Brazil, the government regulates the organic terms for food and cosmetic products. Although the country has national regulations for organic foods, standards for organic cosmetics have yet to be introduced.

10.2.14.2 Asia

There are currently no Asian standards for natural and organic cosmetics. Many Asian countries, such as Thailand, India, China, South Korea and Japan, have national standards for organic foods. It is likely that private standards for natural and organic cosmetics will be introduced in these countries as consumer demand for organic products continues to strengthen. Until then, European standards (especially those of Ecocert Greenlife) will continue to gain popularity.

10.2.14.3 Australasia

The Australian government introduced the National Standard for Organic or Bio-Dynamic Produce in 1992. The standard covers organic foods for the export market and is voluntary for the domestic market. The Australian Quarantine Inspection Service (AQIS) accredited several certification agencies to provide inspection and certification services for the national standard. In July 2009, Standards Australia released their official National Standard for Organic and Bio-Dynamic Produce document that outlines the standards that all manufacturers and producers of Organic Cosmetics and Skincare must adhere to. The organic beauty products standard does appear to be stricter than that of Ecocert and even the Soil Association. Since the Australian national standard is voluntary, very few cosmetic companies are adopting the standard. Natural and organic cosmetic companies, particularly those who export, are likely to continue to embrace internationally recognised standards such as Ecocert.

Australia has several certification agencies for organic foods. The National Association of Sustainable Agriculture, Australia (NASAA), Organic Food Chain (OFC) and Australian Certified Organic (ACO) also have standards for organic cosmetics. The Australian Certified Organic Standard (ACOS) of ACO was published in June 2010. Licensees can utilise the 'Bud' logo, which is promoted as the Australian standard for organic personal care products. The Biological Farmers of Australia (BFA) is behind this drive.

There are no private standards for natural cosmetics in Australia. However, The BFA Natural Ingredients for Cosmetics Scheme allows the evaluation and certification of natural ingredients that are used in ACO organic certified products.

10.2.15 Comparison of the Key Requirements of the Ecocert Greenlife, COSMOS and Natrue Standards

Table 10.1 compares the key requirements of the three major European standards for natural and organic standards [3].

10.3 FAIR TRADE LABELS

The commonly agreed definition of fair trade, as set by leading industry organisations, is as follows:

> Fairtrade is a trading partnership, based on dialogue, transparency and respect, that seeks greater equity in international trade. It contributes to sustainable development by offering better trading conditions to, and securing the rights of, marginalised producers and workers, especially in the South. Fairtrade organisations, backed by consumers, are engaged actively in supporting producers, awareness raising and in campaigning for changes in the rules and practice of conventional international trade. Fairtrade products are produced and traded in accordance with these principles wherever possible verified by credible, independent assurance systems.

Table 10.1 Comparison of Ecocert Greenlife, Natrue and COSMOS Standards.

Criterion	Ecocert	Natrue	COSMOS
% of brand range products that must be certified	No requirement	75%	No requirement
Water counted in the natural content calculation?	Yes	No	Yes
Minimum % of organic agricultural ingredients required to receive certification as a **natural** product	5%	0%	0%
Maximum % of permitted synthetic ingredients allowed	5%	Only permitted 'nature identical' preservatives	Only permitted preservatives and chelating agents
Organic product – minimum % by wt. of organic content (based on total product) – excluding soaps	10%	10.5% (for products 'with organic portion') 19% for organic cosmetics	10 or 20% (depending on product type)
Amphoteric surfactants allowed	Yes	No	Yes (Until end of 2014)

Source: Reproduced with permission from Organic Monitor © 2013.

The strategic intent of fair trade is to:

- Deliberately work with marginalised producers and workers in order to help them move from a position of vulnerability to security and economic self-sufficiency.
- To empower producers and workers as stakeholders in their own organisations.
- To actively play a wider role in the global arena to achieve greater equity in international trade.

Fair trade certification is becoming popular with cosmetic companies because of the importance of ethical sourcing. The fair trade concept guarantees a fair price to growers in developing countries. Developed by the Fairtrade Labelling Organisation (FLO), the Fairtrade standard was initially developed for commodity crops, such as coffee and cocoa. The Fairtrade standard has since been extended to a wide range of products that include wine, flowers, fibres and soccer balls. In some countries such as the UK and France, cosmetic products are allowed to be certified and carry the Fairtrade logo if they meet the standards for composite food products. Thus, companies such as Bubble & Balm and Boots are using fair trade ingredients such as cocoa, vanilla and sugar to launch certified fair trade beauty products [4].

The Fairtrade standard greatly limits the palette of cosmetic ingredients available, as the standard is developed for crops grown in the southern hemisphere. New fair trade schemes have been launched and are gaining ground. The IMO Fair for Life and Ecocert Fairtrade standards are becoming widely popular as they provide more flexibility to cosmetic product developers. The FairWild standard covers wild harvested plants, whilst the IBD EcoSocial standard covers ethical sourcing and biodiversity.

10.4 OTHER ECO-LABELS

Eco-labels tend to evaluate a product using a life-cycle approach to ensure that significant environmental impacts – such as the impact on climate change, nature and biodiversity, energy and resource consumption, generation of waste, emissions, pollution, release of hazardous substances – are considered in the development of the product, raw materials extraction, through manufacturing to use and disposal. Details are some of the leading eco-labels are described in this section [5].

10.4.1 Eco Flower – The European Eco-Label

The 'Flower' is the symbol of the European Eco-label. This is a voluntary scheme, designed to encourage businesses to produce more environmentally friendly products and services. It also enables European consumers to easily identify such products. The functioning of the EU Eco-label is set through a Regulation of the European Parliament and of the Council.

The Flower is used throughout the European Union and related countries such as Norway, Liechtenstein and Iceland. Particular reference is made to environmental impacts, especially the impact on climate change, nature and biodiversity, energy and resource consumption, generation of waste, emissions, pollution and release of hazardous substances. Substitution of hazardous substances with safer alternatives is a key concern. Increased durability and reusability of products is also encouraged. The scheme aims to measure the net environmental effect over a product's life cycle. Social and ethical aspects and a reduction in animal testing are also mentioned.

The ecological criteria for the award of the European eco-label to soaps, shampoos and hair conditioners were first laid out in a 2007 Commission decision document. The criteria have been developed to ensure that only around the 10–20% most environmentally friendly products on the market can meet them. However, the take-up of the Eco Flower scheme in personal care has been quite low. In 2012, the European Eco label was found on over 17 000 products, of which 407 were on products in the soaps and shampoos category [6].

10.4.2 Nordic Swan – The Nordic Ecolabel

The Nordic Ecolabel is the official eco-label for the Nordic countries, introduced in 1989 by the Nordic Council of Ministers to encourage sustainable consumerism. It is also a voluntary licence system where a product must be among the least environmentally damaging in its category, but also possess quality and functionality at least on par with other products. It takes into account human health aspects by specifying that the product should not contain harmful substances.

The label is usually valid for 3–5 years, after which the criteria are revised and the company must reapply for a license. It is claimed that 67% of consumers in the Nordic countries (Sweden, Norway, Finland and Denmark) understand the logo. Mainly, small–medium size Nordic cosmetic companies have so far adopted this standard. Once separate, the main criteria for cosmetics and rinse-off products have now been combined. The current criteria apply until 31 March 2014. Many of the criteria – such as the allowable preservatives – simply state that companies must follow the EU Cosmetics Directive, but with some additional criteria regarding bioaccumulation, endocrine disruption, sensitising effect and ensuring there is no release of Carcinogenic Mutagenic or Toxic to Reproduction (CMR) substances.

10.4.3 Others in Europe

Other schemes include the German Blue Angel mark. Products must meet certain environmental criteria determined by the German authorities, environmental and consumer organisations, and other relevant experts based on a 'cradle to grave' assessment.

In the packaging field, the Program for the Endorsement of Forest Certification Scheme (PEFC) Label indicates sustainable forest management and traceability. Similarly, the Forest Stewardship Council (FSC) is an international non-profit labelling for wood and paper. Its logo guarantees that forests are not over-harvested, animals and plants are protected and that the people who work in the forests are guaranteed training, safety equipment and proper wages.

10.4.4 Green Seal USA

Based in the US, this eco-label has developed a wide range of standards based on ISO standards for environmental labelling programmes. Product or service categories are evaluated using a life-cycle approach to ensure that all significant environmental impacts are considered in the development of a standard: from raw materials extraction through manufacturing to use and disposal. In the personal care field, GS-44 Consumer Soaps Draft Final Standard was finalised in May 2009. This covers soaps, cleansers and shower products. This was followed by GS-50 Green Standard for Personal Care and Cosmetic Products issued April 22, 2011 (first Edition) for leave-on products.

Manufacturing sustainability requirements include use of Good Manufacturing Practice (GMP), reporting of energy usage, emissions and waste.

The standard claims to be the first 'that comprehensively addresses the health, environmental, and labelling concerns for these personal care products'. For example, the standard 'prohibits carcinogens, such as 1,4-dioxane, and other hazardous materials commonly found in these products'. Green Seal requires labelling of ingredients for all products and defines common claims made on these products, such as natural, organic, and biobased. Organic claims must be supported with documentation that they meet the USDA National Organic Program or the NSF ANSI 305 standard.

10.4.5 USDA Biobased Product Certification

The USDA BioPreferred® program for products made with 'renewable biological material' is a wide-ranging scheme covering many industries [7]. The 'USDA Certified Biobased Product' seal lists how much of the product is made from corn, soy or other renewable sources. It is designed to be a consumer seal that is becoming evident on store shelves; however, it includes only a few personal care and toiletry products.

There is a required minimum biobased content based on the amount of qualifying biobased carbon as a percentage of the weight of total organic carbon in the finished product. This percentage is determined by product type. For example, a shampoo would be expected to meet a minimum 66% biobased content and skin care lotions and moisturisers the proposal is to meet 59% minimum biobased content.

10.4.6 Carbon Labels

Growing concerns about climate change is leading organisations and consumers to look at the concept of 'carbon footprint': the totality of all the carbon dioxide emissions that result from any particular product, service or human activity. Companies are finding new ways to measure and ameliorate the total environmental impact of their products. Companies looking to get their products carbon labelled have several options for carbon labels.

In Europe, several retailers have introduced voluntary carbon labelling: the French supermarket chains Casino and E. Leclerc, Switzerland's leading supermarket chain Migros, and the UK's Tesco. A small but growing number of cosmetics companies have attained carbon neutral certification. The Brazilian company Natura Cosméticos and UK-based Neal's Yard Remedies have declared themselves carbon neutral. Smaller, typically natural and organic, cosmetic companies are also becoming carbon neutral. Perfect Organics, an American brand, announced in 2007 that it had offset its carbon emissions with Carbonfund.org. New Zealand-based Snowberry has achieved carbon neutral certification for all its luxury skin care products. Snowberry's certification has been administered by Landcare Research's carboNZero.

Weleda is working with Soil and More to become carbon neutral. L'Oréal is working with retailers, including Tesco and Wal-Mart, on a single methodology to carbon-label its products. The French beauty firm has stated that it prefers to use one universal label to display the carbon footprint of its products, however no such label is yet recognised worldwide. L'Oréal has had to concede that it may need to use several labels.

10.5 OTHER SUSTAINABILITY STANDARDS AND INDICES

10.5.1 ISO Standards 14000 and 26000

The International Organisation for Standardisation (ISO) is the world's largest developer of international standards. Founded in 1947, it now has a membership of 163 national standards institutes. ISOs portfolio of more than 19 000 standards provides practical tools for all three dimensions of sustainable development: economic, environmental and social. Published under the designation of International Standards, ISO standards represent an international consensus on the state of the art in technology or best practices. The two main standards relevant to sustainability are ISO 14000 and ISO 26000 [8].

The ISO 14000 family of standards for environmental management was launched in 2004 to provide a practical toolbox to assist in the implementation of actions supportive to sustainable development. It reflects international consensus on good environmental and business practice that can be applied by organisations all over the world in their specific context. ISO 14001 is the world's most recognised framework for environmental management systems (EMS) that helps organisations both to manage better the impact of their activities on the environment and to demonstrate sound environmental management. It offers guidance for introducing and adopting environmental management systems based on the best universal practices, in the same way that the ISO 9000 series, which is now widely applied, represents a tool of the best available quality management practices. ISO 14001 has been adopted as a national standard by more than half of the 160 national members of ISO and its use is encouraged by governments around the world. Although certification of conformity to the standard is not a requirement of ISO 14001, at least 223 149 organisations in 159 countries and economies were using it in 2010. In order for an organisation to be awarded an ISO 14001 certificate, it must be externally audited by an accredited audit body.

The ISO 26000 standard for social responsibility was released in November 2010 after five years of negotiations between governments, NGOs, industry representatives, consumer groups and labour organisations. ISO states it represents an international consensus standard.

Known as ISO SR, the standard does not contain requirements and therefore, in contrast to ISO management system standards, is not certifiable. The ISO 26000 scope clearly states

> This International Standard is not a management system standard. It is not intended or appropriate for certification purposes or regulatory or contractual use. Any offer to certify, or claims to be certified, to ISO 26000 would be a misrepresentation of the intent and purpose and a misuse of this International Standard. As this International Standard does not contain requirements, any such certification would not be a demonstration of conformity with this International Standard.

This ISO 26000 scope also includes that the international standard cannot be used as basis for audits, conformity tests and certificates, or for compliance statements. As a guidance document, the ISO 26000 is an offer, voluntary in use, and encourages organisations to discuss their social responsibility issues and possible actions with relevant stakeholders. The standard encourages to organizations to reconsider social responsibility or 'socially responsible behaviour' and to identify/select from its recommendations those where the organisation could/should engage in contributions to society. ISO 26000 encourages further to report on actions taken.

A critique of the standard is that it can have low practical value, since it provides a common understanding of social responsibility instead of also facilitating management routines and practices leading to social responsibility.

France-based Laboratories Expanscience is one of the few cosmetic companies to adopt the ISO 26000 standard. The company has integrated the principles of the social responsibility standard into its supply chains for vegetal ingredients [9].

10.5.2 SA8000

SA8000 is a global social accountability standard for decent working conditions, developed and overseen by New York-based Social Accountability International (SAI). SAI contracts with a global accreditation agency Social Accountability Accreditation Services (SAAS), which licences and oversees auditing organisations to award certification to employers that comply with SA8000. SA8000 is based on the UN Universal Declaration of Human Rights, Convention on the Rights of the Child and various International Labour Organization (ILO) conventions. SA8000 covers areas such as child labour, forced labour, workplace safety and health, freedom of association and right to collective bargaining and so on. Pressed by large cosmetic companies, a number of cosmetics contract manufacturers, especially those with factories in Asia, have obtained SA8000 certification

10.5.3 Other Standards

AccountAbility is a London-based independent, not-for-profit organisation promoting accountability, sustainable business practices and corporate responsibility. It is a self-managed partnership, governed by its multi-stakeholder network of leading business, public and civil institutions.

Its **AA1000 standard** series help organisations become more accountable, responsible and sustainable. The series comprises open source frameworks developed through a multi-stakeholder consultation and review process. The standards are designed to be compatible with other key standards in this area, including the GRI Guidelines, SIGMA Guidelines, SA8000, the ISO Series and financial accounting standards.

The AccountAbility Principles Standard (AA1000APS) provides a framework for an organisation to identify, prioritise and respond to its sustainability challenges. The other AA1000 standards – the Assurance Standard (AA1000AS) and the Stakeholder Engagement Standard (AA1000SES) – are based on the APS principles and support their achievement.

The **GoodCorporation Standard** was developed by London-based GoodCorporation in cooperation with the Institute of Business Ethics in 2000. The second revision was introduced in 2007 following a public consultation. Unlike previous initiatives that looked at specific aspects of management or tackled industry sector issues, the GoodCorporation Standard claims to look at all the key relationships and impacts in a consistent way.

The GoodCorporation Standard is also an open source document. Organisations are encouraged to download and use the standard to improve and develop their own Codes of Conduct. The standard sets out 62 management practices that are individually assessed. Those that meet the standard qualify as GoodCorporation members. L'Oréal is the only major cosmetics company with GoodCorporation accreditation.

10.5.4 Sustainable Indexes

A financial index is a statistical measure of the value of a certain portfolio of investment vehicles. Sustainable financial indexes serve as a basis for responsible investment, financial instruments and fund products, providing asset managers with reliable and objective benchmarks to manage sustainability portfolios. They are also useful as:

- A reference tool to provide companies with a transparent and evolving global corporate responsibility standard.
- A research tool to identify environmentally and socially responsible companies and select adequate business partners and suppliers.
- A benchmark index to track the performance of competitors.

The FTSE4Good Index Series and Dow Jones Sustainability Indexes are the main global indexes tracking the financial performance of the foremost sustainability-driven companies worldwide.

Launched in 1999, the Dow Jones Sustainability Indexes (DJSI) are a family of indexes evaluating the sustainability performance of the largest 2500 companies listed on the Dow Jones Global Total Stock Market Index. They are the longest-running global sustainability benchmarks worldwide, and have become the key reference point in sustainability investing for investors and companies alike. The DJSI is managed by S&P Dow Jones Indices and Sustainable Asset Management (SAM).

Launched in 2001 by the Financial Times and Stock Exchange (FTSE), the FTSE4Good Index series is a series of ethical investment stock market indices. The series has been designed to objectively measure the performance of companies that meet globally recognised corporate responsibility standards. Research for the indices is undertaken by FTSE and the Ethical Investment Research Services.

In the UK, the **Corporate Responsibility (CR) Index** is the leading voluntary benchmark of corporate responsibility. It has been implemented by the charity Business in the Community since 2002.

A good example of a company included in these indexes is Unilever. For over 10 years it has been a top 20 company in the Dow Jones Sustainability World Indexes. It retained its Platinum Plus standard in the UK's Business in the Community CR Index 2012, and was included again in the FTSE4Good Index Series.

Many countries have standards for sustainability and/or Corporate Social Responsibility (CSR). Some, like the British Standard for Sustainable Development (BSI 8900:2006), cover various aspects of sustainability, whilst most others cover specific areas like recycling, packaging and environmental management. No country has introduced a specific sustainability standard for the cosmetics industry.

REFERENCES

[1] Organic Monitor, (2011) Global Market for Natural & Organic Cosmetics, London, UK.
[2] International Organic and Natural Cosmetics Corporation http://www.ionc.info (accessed on June 2012).

[3] Organic Monitor (2009) Technical Insights Report on Critical Review of Natural & Organic Cosmetic Standards, London, UK.
[4] Organic Monitor (2008) Strategic Insights report on The Potential of Fair Trade Cosmetics & Ingredients, London, UK.
[5] Organic Monitor (2010) Strategic Insights report on CSR & Sustainability in the Cosmetics Industry, London, UK.
[6] European Commission EU Ecolabel http://ec.europa.eu/environment/ecolabel/ (accessed on June 2012).
[7] USDA Biobased Preferred Program http://www.biopreferred.gov (accessed on June 2012).
[8] International Organization for Standardization http://www.iso.org/ (accessed on June 2012).
[9] Sustainable Cosmetics Summit Europe, Paris, November 21–23 2012.

11

Understanding Green Marketing

Darrin C. Duber-Smith and Mason W. Rubin

This product is Natural! This company is Socially Responsible! Their product isn't Environmentally-Friendly! But our business efforts embrace Sustainability! We are Greener than thou! Even the casual observer is no stranger to the fact that these concepts have become integral parts of organizational operations and marketing strategies. Sustainability, the most current and all-encompassing term, has been defined by these authors as an organization's effort to meet corporate objectives and consumer needs in a way that demonstrates continuous improvement toward minimizing negative impact on people and the natural environment.

A large number of businesses in every industry have, for a variety of reasons, employed strategies to address the growing concern for human health and the natural environment. But the pioneers can all be found in the natural products industry. Partly a backlash against the "Better Living Through Chemistry" zeitgeist of the early and middle 1900s, the healthy foods movement, which is now an industry worth hundreds of billions of dollars, was also a response to the environmental devastation that was perhaps best illustrated in Rachel Carson's early 1960s blockbuster *Silent Spring*. "Is what makes the Cuyahoga River catch fire also in our food and water supply?" The environmental problems that began to be addressed around this time raised social consciousness beyond that of simply focusing on the steamroller of economic development that occurred after World War II. Was it worth having all of these technological advancements at the cost of the ecosystem's health? Apparently not.

Sustainability: How the Cosmetics Industry is Greening up, First Edition. Edited by Amarjit Sahota.

The term "Green Marketing" is a colloquial one. It may have been coined by Jacqueline Ottman, author of the book *Green Marketing* in the early 1980s. The term also appeared in Dr. Philip Kotler's famous *Introduction to Marketing* textbook around that time, so it is difficult to say to whom the term's origination should be attributed. These terms quickly infiltrated both industry and academic circles, and it is difficult to assign derivation to such colloquial words. Nevertheless, the term adequately described a company's efforts to address what is more scientifically termed, "Sustainability and Corporate Social Responsibility," and unfortunately the word has been used liberally to describe all manner of commitment (or lack thereof) to sustainability efforts.

With its start in small "hippie" health food stores and "pill shops," the natural and organic products industry began to boom in the late 1980s, driven by the growth of the nutritional supplement category and the beginnings of mass distribution. In the USA, the Dietary Supplement Health and Education Act of 1994 insured that vitamins, herbs, and all manner of supplements remained available to consumers over-the-counter, and also outlined a list regulatory requirements that remains in the process of development even as of this writing. Soon the functional foods area began to grow, as an increasing number of consumers began to prefer their supplements in food or beverage form rather than a pill, tincture, or powder. The profound growth of natural and certified organic foods and beverages of all kinds followed, and most recently, the growth has been in the personal care and household products sector (global sales reaching USD 8.4 billion in 2010 according to Organic Monitor) as positive attitudes toward health and the environment have become increasingly pervasive.

The double-digit growth of the natural products industry over a period of many decades has influenced companies in many other categories of consumer and business goods to address this social trend with regard to other types of goods and services. From raw materials to the end user, sustainability is now part of the process. It is true that consumers are fickle, and that current social attitudes toward sustainability may take a back seat to concerns such as terrorism and the economy, as we have seen recently; but the practices that have been integrated into industries at all levels cannot be easily changed if efforts to be more sustainable have been made in good faith. As we will see later, sometimes this is not the case.

11.1 THE "WHY" OF SUSTAINABILITY

Indeed, there is some evidence that the efforts of many businesses, at least in the realm of communications, have gone unheeded. Research published in the January/February 2011 edition of *Nutraceuticals World* sheds light on the failure of organizations to effectively communicate sustainability initiatives. According to the Hartman Group, even though 15% more consumers are aware of the term "sustainability" versus three years ago (a total of 69% are now aware of the term), only 21% can readily identify a sustainable product and an even more disturbing just 12% can name companies that are sustainable. Obviously what we have here is a failure to communicate, which begs the question as to why organizations would engage in sustainability initiatives if they do not plan on communicating them effectively? First, we will address the drivers behind sustainability initiatives and move to how such a program can be initiated. Sustainability plans are developed and implemented for the following reasons, coined by Duber-Smith as the "Green Imperative":

1. **Target Marketing:** A sustainable marketing strategy, with products that are properly positioned, will address the growing target market for goods that are green in some way, whether natural, certified organic, recyclable, made from recycled materials, dematerialized, and so on. Numerous market segmentation studies have identified a growing market for greener goods and services among the majority of Americans.
2. **Sustainability of Resources:** Ensuring the availability of resources to continue to make and sell goods is another imperative that suppliers, manufacturers, and retailers must embrace. Cutting down all of the trees does not help the shareholders of paper companies, let alone everyone else.
3. **Lowered Costs/Increased Efficiency:** There are countless ways to save money and increase efficiency so that marketers can enhance the bottom line and stave off the narrowing of margins that occurs in every industry as it reaches the maturity stage of the life cycle.
4. **Product Differentiation and Competitive Advantage:** Every marketer knows that in this hyper-competitive business environment it is crucial to maintain demonstrable advantages over competitive and substitute offerings. Green products are often a "tie-breaker" when consumers are faced with a "greener" product offering the same benefits as a product perceived to be less green.
5. **Competitive and Supply Chain Pressures:** When competitive organizations and their products adopt sustainable business models and green positioning, it often pressures other companies to follow suit, especially in the case of market leaders. Wal-Mart and its recent environmental and social initiatives illustrate how powerful supply chain members can force companies around the world to adopt more favorable social and environmental policies.
6. **Regulation and Risk:** Regulations at all levels of government are on the rapid rise, so organizations need to not only remain in compliance but also proactive with regard to impending legislation. This practice reduces shareholder risk.
7. **Other Stakeholder Demands:** Activist shareholders, NGOs, the financial sector, and the media all work independently and sometimes in concert to ensure that companies are cognizant of their impact on people and the environment.

8. **Brand Reputation:** Any marketer worth his/her salt knows that a brand's reputation is of paramount importance, and being sustainable enhances that reputation among the majority of stakeholders.
9. **Global Market Forces:** Global concerns about climate change, looming energy problems, and a recent growing backlash against globalization among many others factors all point toward the necessity in addressing sustainability issues.
10. **Customer Loyalty:** A brand's attitude toward sustainability is just one of the many variables that factor into the decision-making processes of the majority of consumers.
11. **Employee Morale:** A wide body of research points to the fact that adopting a more sustainable business model actually enhances employee morale.
12. **The Ethical Imperative:** This concept is simple. It is not ethical to degrade the environment and the people in it in the name of commerce. Embracing sustainability is simply the right thing to do, and stakeholders are sensitive to this.

A careful examination of the above reasons for building a sustainability model into your business and marketing strategy reveals these efforts should lead to the magic words, "Return on Investment." This is simply good business.

11.2 THE GREEN CONSUMER

The authors embrace "the marketing concept," that is a marketer assesses a consumer need FIRST, and then makes a product to meet that need. Even the most amazing of innovations meets a basic need, from the smart phone meeting a need for non face-to-face communication to the motor vehicle continuing to meet the need for non-human powered transportation. Many product failures result from creating an innovation and THEN finding a need in the marketplace. This was a common practice in a much simpler time, before there were some 600 different kinds of toothpaste on the market.

The needs met by food, supplement, personal care, household, and other products are obvious and the fact that they may be better for people and the environment is not only a bonus, but also increasingly a driver. As there are many marketing research companies that study the natural and organic product industry's trends, there are a number that specialize in tracking consumer demographic (describing consumers in terms of size and structure), geographic location, and psychographic (behaviors and attitudes) trends. One of the most prolific of such studies is the LOHAS Segmentation (as described below). This study is developed and fielded at least annually by the Natural Marketing Institute (NMI) and now has over 12 years of history in identifying and tracking consumer preferences and characteristics through market segmentation studies.

LOHAS stands for Lifestyles of Health and Sustainability, and the study is commissioned primarily to identify appropriate target markets for natural, organic, and environmentally friendly goods and services. Natural personal care represents one of the fastest growing segments of this marketplace. The Natural Marketing Institute considers a natural personal care user to be one that has purchased at least one product during the previous six months. In 2009, according to NMI, 30% of general population consumers classified themselves this way. This number has surely grown since then.

The LOHAS segmentation divides the US general population into five market segment categories based on demographic, geographic, and psychographic characteristics reflecting attitudes and behaviors with regard to natural products.

The primary target for most natural products is the aptly named **LOHAS segment**, consisting of about one-fifth of the general population. Consumers in this segment tend to live health as a lifestyle and will go out of their way to purchase natural and organic products. They are early adopters, opinion leaders, and avid users of green products. They aren't necessarily high income and tend to be higher educated. While concentrated more heavily on the coasts, they are actually located throughout the USA.

Naturalites, another attractive market segment, representing about another one-fifth of the general consumer population, are more interested in health than they are in the environment; and they are primary users of natural and organic personal care. They are higher income and tend to be better educated than the general population.

Drifters, the largest market segment at about a quarter of the general population, tend to be much younger and have yet to develop their own value structures and engrained purchase patterns. They are motivated by the latest trends and shift on commitments to many issues, but it is probable, because of current attitudes and behavior, that the majority of these drifters will "drift" to the LOHAS or Naturalites categories as they get older.

The other two segments, **Conventionals** (lower income folks motivated by price and practicality) and **Unconcerned** (the nomenclature reflects the general attitude) are not primary targets for greener personal care or household products unless they can save them money, but the ongoing proliferation of natural/organic cosmetic product availability across multiple channels of distribution, coupled with the growing health and wellness market, might change this. These segments are ignored in the targeting process for now, but must still be monitored as the growing popularity and availability of natural products will surely bring some of these consumers from these two segments into one of the other three segments. The bottom line is that well over half of adults are in the primary target for natural and organic personal care products through the LOHAS segmentation model.

Again, this is only one example of many studies dedicated to studying this field; but the author's own personal bias, coupled with the consistent findings of the study over a long period of time, make it a very appropriate example for this book.

11.3 BEST GREEN PRACTICES

Much of marketing is exaggeration, a common practice sometimes known as "puffery." This, and the utter proliferation of marketing messages over the past several decades, have taken its toll on the consumer. We are bombarded with thousands of messages each day, and our filters are becoming rather effective at message selection. This is why so many people are still skeptical about industry's commitment to greener business practices, and still too many downplay the efforts of the most angelic of companies. Still more are unaware of the company's efforts in the first place. Others are left to sift through the marketing clutter to assess which organizations are the real deal, and which ones are engaging in the odious practice of "greenwashing."

In the name of complete transparency, the best way for a company to employ a green marketing strategy is for a sustainability audit and plan to be available on a company's web site and updated annually with measurable objectives for improvement in the following areas:

- The nature of raw materials and composition of products offered.
- The nature, consumption, and recapture of energy.
- The use of water.
- Impact on land and biodiversity.
- Reduction and recovery of emissions, effluents, and waste.
- Distribution issues such as packaging and transportation.
- Cause related involvement.
- Human resources and vendor partner policies.

Commitments to any of these areas can be easily incorporated into a product's brand identity and crafted into a message that embodies a concern for people and the natural environment. These days there are numerous models to follow, as well as a variety of experts one can employ to ensure objectivity, thoroughness, and an adequate level of expertise. Conducting an audit across the factors listed above, setting measurable objectives for continuous environmental and social improvement, and monitoring the results is the best way to build a foundation for communicating authenticity.

The most influential global model for sustainability planning is probably the Global Reporting Initiative (GRI). In 1997, the GRI was launched by Boston-based CERES, an international network of investors, environmental groups, and other public interest organizations formed to address sustainability challenges. The program produces the world's most widely used reporting framework to maximize comprehensiveness and transparency. As such, GRI has become the de facto international standard used by well over 1000 companies for corporate reporting on environmental, social, and economic performance.

For the "uber green" crowd (and especially if the company involves a global brand), it may be a good idea to obtain global certification. Those in manufacturing may be familiar with the ISO (International Standards Organization) 9000 series of global certification for quality management and may also recognize ISO-14000. This series exists to help organizations minimize how their operations negatively affect the environment – that is, compliance with applicable laws, regulations and other environmentally-oriented requirements, as well as how to continually improve across these areas. ISO-14000 is similar to ISO-9000 in that they both pertain to the process of how a product is produced. Obviously, adding a natural or certified organic element to the product is par for the course.

11.4 COMMUNICATION VERSUS REALITY: THE MANY SHADES OF GREEN

So, ultimately, what does it mean to be green, and what do some of these organizations look like? There are many shades, from outright "greenwashing" (a largely fraudulent effort to position a brand as environmentally and socially responsible) to bona fide attempts at comprehensive reform by organizations to be "as green as they can be" (without compromising shareholder value, of course).

The ethical implications of all of this should be fairly obvious to the observer, but it does not matter so much what the company's actual green practices are, as much as it matters that marketing communication is transparent and truthful. Bearing this in mind, the authors have identified six archetypes, ranging from not-so-green to green-as-it-can-be, so that the reader may gain a better understanding as to how sustainability initiatives are being implemented and communicated in the marketplace.

A trend that is discussed in the following archetypes is that the cosmetics industry as a whole is becoming greener. More and more companies are implementing sustainability plans and adding environmental impact goals to their company mission statements. This trend mirrors that of the larger green movement in the Western world as a whole. Consumers are becoming increasingly sensitive to how green companies are and this can be a differentiating point of competitive advantage that many companies are starting to take advantage of. This trend is becoming increasingly pervasive in the cosmetics industry especially; expect to see this trend accentuate in the years to come as governments pass more sustainability laws and consumers become more aware of how products affect the environment.

11.4.1 Red Marketer

According to the color spectrum, red is green's opposite color. Red marketers are those who absolutely ignore all of the principles discussed earlier. These firms are becoming increasingly rare in the industrialized world, but still exist in great numbers in the developing world. Companies in this category have no sustainability efforts whatsoever to communicate and have no plans to develop any such efforts. There are few, if any, cosmetic brands in this category, but there may be several suppliers of synthetic chemicals that have failed to address social and/or environmental issues thus far.

Red marketers are becoming an extinct breed among first world and Western cosmetics companies. Companies have seen how important it is that they show at least some green initiative, even if it is just that – show. Almost all major cosmetics companies in the United States and Europe have jumped on the bandwagon and have some sort of sustainability plan or green action that they support. Not to do so would mean bad publicity as well as lost brand equity in the mind of consumers. Over the last decade, as green business practices and sustainability have become the standard to which companies are held, companies in all industries have started to "green up" in an effort to quell bad press, stay competitive in the marketplace, and pre-empt lawsuits and future industry regulations. This phenomenon is especially prevalent in the cosmetics industry, because consumers feel that products that they ingest or put on their bodies are more likely to do them harm, so they should be more natural and green-inclined in those areas. Thus, there has been more of an outcry in recent years for cosmetics companies to show sustainability and use natural ingredients than in many other industries.

The only place that true red marketers exist in is countries that are still developing or that do not have strong laws or values towards sustainable business practices. Some companies in Asia fall into this category, primarily because there is little incentive to become green in their respective countries because there are few laws or regulations that force them to be. Additionally, consumers are less concerned with sustainability in many of these areas, so it is not a relevant point of differentiation between products. Because of this, many companies focus on making their products as efficiently and inexpensively as possible and little else. The green imperative is just not as important or pertinent to these companies as it is to companies and consumers in the United States and Europe.

Bodylife Cosmetics is a company based out of Guangzhou, China, that produces a vast array of different cosmetic products. The closest thing this company has to a green business practice or sustainability plan is a clean workshop examination certificate. The company focuses on what it does best – producing products in large quantities efficiently and inexpensively. There is no attempt to engage in sustainable business practices; the only objective is maximum return on investment. Although red marketers do not exist in American and European cosmetics companies anymore, they are still prevalent in other parts of the world. Sustainability is a trend that is becoming more and more global; it has already started to permeate most foreign markets because of the pressures of global commerce. The age of the red marketers is almost over.

11.4.2 Green Panderer

A panderer can be defined as someone who serves or caters to the passions or plans of others, often to make money. Sound familiar? Thus organizational panderers are those that pay lip service to the green market with very little to zero commitment of resources to actually achieve any meaningful level of sustainability. This type of company may only talk about green issues to give the impression that it is participating in the solutions because of market pressure or because it sees a temporary opportunity to make money by doing so. These organizations are "greenwashers" to the extent that a thin, "green sheen" can be detected. Such an organization may support a cause or reduce energy consumption in some minor way that resonates with stakeholders, but that will be the extent of the commitment. Many larger major manufacturers and ingredient suppliers that entered the natural and organic products space several years ago were accused of pandering. Simply offering one or two natural ingredients among a sea of unsustainable business practices is a thin sheen indeed.

As more and more companies join the "green movement" and start to develop sustainability plans and advertise their green products or business practices, it is easier to spot which ones are actually green and sustainable, and which ones are merely pandering their "green" practices in an attempt to appear green to consumers. As an overall trend in the industry, most companies are implementing sustainable and green practice so these green panderers are few in number. Most companies see the necessity of sustainable and green business in today's ever-changing global market. However, there are some companies out there who haven't really made the effort to "green up" yet. One of these companies is France-based Pierre-Fabre, which offers a variety of dermo-cosmetics.

Pierre-Fabre does not have a sustainability plan, their products are not natural, and they do not have any sort of green-inclined goals. They put a lot of emphasis on how their products are made from plant extracts and use some natural products, but this is the only way that they appear to be green. Most of their focus is on how beneficial and high quality their products are. Little or no effort is put into the green imperative.

Companies like Pierre-Fabre try to show that they are green in some small way, but overall they are not being sustainable at all, and that is why they would be categorized as a green panderer. They are trying to appear green and pander that image to their consumers. This is greenwashing at its finest and should be scrutinized as such. The cosmetics industry is moving toward greener practices and sustainability as a whole. Companies that merely continue with the same business practices as they have been using for decades are going to lose favor among consumers as more and more people become sensitive to greenness as a positive product attribute. Companies like Pierre-Fabre need to rethink their sustainable business practices if they are to survive in the modern global marketplace.

11.4.3 Green Buffeteer

This classification of organization refers to those that choose the easiest and most obvious greening options with little attempt at transparency or commitment to continuous improvement. This rather shallow effort is slightly more than that shown by the green panderer crowd, but still only penetrates skin deep into business practices, thus creating a "green skin." Many of the efforts with regard to these organizations are socially responsive or seen to be profitable in the short or long terms. Increasingly green-sensitive consumers can easily see past this skin into the true heart of the organization and will not be fooled for long. Recycling, conserving water, and reducing energy are all relatively easy and visible ways to achieve a modicum of sustainability without a significant commitment, and the green buffeteer rarely sets or sticks to measurable objectives making it difficult to assess results.

The cosmetics brand Herbal Essences is an example of a green buffeter in that they have a sustainability plan. However, it contains only the basic green practices that can give the company a green image on the outside, but in reality contains a few easily implemented changes to their business practices that only make a minor impact. Herbal Essences is part of the large conglomerate Procter & Gamble (P&G), and is thus subservient to their policies on sustainability. P&G has a sustainability plan and has set environmental impact goals, but these are criticized as limited to some of the easiest sustainable practices large corporations commit to. P&G's goals include increases in renewable energy and packaging and conserving resources. Their goals set for 2020 include reducing their petroleum-based packaging by 25%, and increasing their renewable energy consumption by 30%. These goals are all well and good, but there is a lot more the company could be doing. The company appears to be supporting a few sustainable practices that look good on paper rather than engaging in the other more difficult sustainable practices available to them. For instance, they could make commitments to the ethical sourcing of raw materials, offsetting carbon emissions, and using green formulations. P&G's "long term vision" includes emitting no fossil-fuel based emissions, using 100% renewable packaging, and having their plants completely powered by renewable energy. They have not placed any timeline on these goals, so it is easy to see how they could fall by the wayside over the years.

Another example of how Herbal Essences is a green buffeteer is shown through their reformulation of their shampoos in 2010 in order to reduce toxins. This would be a step in the right direction for the company if it was undertaken of their own accord. The Green Patriot Working Group was in the process of filing a lawsuit against Procter & Gamble, for violating Proposition 65 limiting cancer-causing dioxane levels in consumer products, when they announced that they would be making a new formula that met industry standards. This suggests that Procter & Gamble only does what it takes to appease the green community and that they are possibly committed to being sustainable. Many companies in today's marketplace fall under the green buffeter category because it is relatively easy to implement and it shows that the company cares about the environment. Unfortunately, most consumers think that this is good enough and these companies truly care about the environment, when in reality they are usually only doing what it takes to avoid chastisement from the green community.

11.4.4 Light Green Marketer

These organizations have strategies to encourage consumers to see a brand that meets or exceeds all of their expectations. These strategies will normally satisfy many consumers, even those who are moderately green sensitive. Motivations for companies to adopt a light green approach range from market pressures, desire for higher profits, and crafting a more socially responsible image. Efforts by these organizations transcend the green buffeteer's attempts at addressing only the tastiest morsels, in that more areas are addressed, suggesting a greater commitment to the effort. The organization sets measurable objectives across a few areas and endeavors to achieve those objectives. May Kay Cosmetics might be a good example here, with a focus on a few high-profile women's causes and other issues, but little to no attempt at sustainable and socially responsible business practices, such as the inclusion of safer, healthier product offerings.

L'Oréal is one of the few large cosmetics companies that is going above and beyond what the market asks of it and has truly committed to moving toward sustainable business practices and reducing environmental impact. In 2011 L'Oréal was named among the top green companies in the world, and most notably in the cosmetics industry. Their sustainability plan is very comprehensive and doesn't only focus on a few easy-to-implement green practices. Some of the notable green practices that L'Oréal has executed include reduction of waste, greenhouse gas production, and water consumption, package re-engineering, production efficiency, and product innovation. L'Oréal goes further than most companies in all of these areas and sets and meets goals regularly.

Between 2005 and 2011, L'Oréal reduced their greenhouse gas emissions by 30%, reduced water consumption by 23%, and reduced transportation waste by 24%. These met goals that they had set previously and they have new ambitious goals for 2015 for having all of the numbers reach 50% reductions in their respective categories. Ninety percent of L'Oréal's factories have received the ISO 14000 environment certification. Many of its factories are powered partially by renewable energy and they have invested millions of dollars into solar panels and wind turbines among other types of renewable energy. They also use delivery trucks that are completely electronically powered in many parts of the world. Waste reduction is important to them as well – they reuse, recycle, or recover 96% of the waste from energy production and over 50% of their plants send zero waste to landfills because of their strong recycling practices. These are the types of actions that make L'Oréal a light green marketer. They have pledged an immense amount of time, effort, and money to the proliferation of the green imperative that shows that they have a substantial commitment to the environment and sustainability. Yet many brands lack natural, certified organic, fair trade, and other types of the more "benign" ingredients embraced by the natural green marketer.

11.4.5 Natural Green Marketer

Most organizations in this category are operating in the natural and organic products industry. They are committed to making their products as green as possible, with a focus on the environmentally and socially benign use of resources, processing methods, and end products. However, these philosophies and practices are somehow not always applied to the organization as a whole. Terms such as natural, organic, and fair trade are commonly used in marketing communications. These companies are willing to sacrifice some short-term profitability for long-term reputation and sustainability of the resources needed to operate. Motivations for the natural green strategy begin to turn more toward inherent social responsibility values rather than external forces. Such an organization might make 100% natural products, but has not adequately addressed the bigger organizational picture, such as what supply chain members and non-product aspects of the organization are doing.

One of the largest natural green marketers in the cosmetics industry is Jason Natural Products, a brand owned by natural products conglomerate Hain-Celestial Group. They have completely natural products and use ingredients that are 70% organic. However, they do not have a sustainability plan or any sort of visible green business practices. Having all natural products limits the types of toxic waste that the company produces, however there are a host of other green areas that the company could focus on that they have instead ignored. It is not enough anymore to merely have natural or organic products. In order to be a true green marketer, a company must couple that with an extensive sustainability plan that demonstrates their dedication to the environment and the green imperative. Since Jason's inception in 1959, they have always advertised their products as everyday natural cosmetics that are good for you and do their job in a natural, chemical-free way. Since then, not much has changed, in that they are still a natural products company, but they have not joined the green movement in sustainability. This puts them square in the natural green marketer category because they are definitely an all-natural products company, but are extremely lacking in other areas of the green imperative.

11.4.6 Deep Green Marketer

These organizations are blazing new paths beyond a new frontier. Their strategies afford the most opportunity for organizations to completely embrace The Green Imperative and infuse sustainability and social responsibility into all of their products, processes, and policies (without negatively effecting long-term shareholder value). These organizations will be the ones that reap the long-term benefits of sustainability, reducing costs, and expanding the market for their goods and services. Absolute transparency and an unwavering commitment to both people and the natural environment are musts. No process or product or partner can be immune from these companies' efforts to maximize efficiency and appeal to the global shift toward environmental and social responsibility. The primary motivation for such a strategy is engrained in the mission statement and is therefore meant to be a constraining factor in terms of organizational strategy. A sustainable company can't stray from the course. Doing well by doing good will ensure that an organization reaches its return on investment (ROI) objectives and that these objectives are met without compromising the ability for future generations to meet their objectives.

The deep green marketer embodies everything that we want to be moving toward in sustainability. They are the trend setters that many in the industry try to follow. Pangea Organics is one of these companies, based out of Boulder, Colorado (a haven for the natural products industry). It is at the forefront of natural and organic products, sustainable operations, and innovative packaging. Their values penetrate all of their business practices and are exemplified in their triple-bottom line of people, profit, and planet.

Pangea Organics is adamant that integrity and honesty are among their core values, so all of their business practices are transparent, as are all of their ingredients and packaging materials. All of their cosmetics are certified USDA organic and are free of petroleum-based substances, sulfates, detergents, synthetic preservatives, and GMOs. Their packaging is completely recyclable, compostable, and biodegradable. Their products are also biodegradable and are not harmful in any way to the environment. Their factory is almost completely sustainable and has close to zero waste. Their carpets are made of 100% post-consumer products and the paint on their walls is VOC free. They reuse packing peanuts instead of throwing them out, and use boxes to ship products that are made of completely post-consumer recycled material. Their manufacturing plant and office is powered by 100% wind energy and they have very low levels of greenhouse gas emissions.

As can be seen, the green imperative is deeply engrained in every aspect of business for Pangea Organics. They are doing everything possible to have as little effect on the environment while providing a quality organic product to its customers. This is the embodiment of a green company that the cosmetics industry is moving toward. Some companies are obviously a lot closer to reaching this level of sustainability than others, and of course some companies are more concerned with becoming green than others. But it is undeniable that becoming greener is an increasingly pervasive trend in business as a whole, but especially in the cosmetics industry.

11.5 GREENER THAN THOU

Unfortunately, one of the major barriers for companies in becoming a natural or deep green marketer is the short-term earnings mentality of many investors, particularly with regard to the publically traded companies, which must meet expectations for sales and profits on a very regular basis. Privately held organizations have more flexibility to make longer-term investments in sustainability, allowing for future returns on investment, in much the same way as investing in research and development. So, what does the future hold for green business?

The global natural personal care category, for example, grew at a 8% clip in 2010 according to Organic Monitor, a leading industry observer; this healthy growth will probably continue as it did for a number of other natural products categories over the past several decades. This rate is down from many years of double-digit growth, but the product category is now large enough that maintaining a high growth rate becomes more difficult. The larger and more mature the industry is, the more the industry needs new consumers to enter it in order to maintain healthy growth rates. This industry is maturing, but is still at the tail end of what we would consider to be a "high growth" stage.

What about the economy? For certain, the green priorities many people may have during good times, can fall out of favor when larger worries like unemployment and low income growth take precedence. We have seen that in an economic downturn, most people do not abandon natural and organic products, but rather decrease their usage, trade down to lower price point brands, and/or purchase products that are on deal. This is especially true for mid-level brands, as consumers are more likely to trade down from a mid-level brand to a lower brand than they are trading down from a premium brand to a low-level brand. The recent performance of Tide, Cheer, and Gain over the past few years illustrates this perfectly.

The days of charging a premium price for a greener product just because it is greener may be at a close. The sheer volume of products, expanding channels of distribution, and the growing target market all point to one thing ... price pressure. Producers and retailers of premium goods will have to demonstrate that their products are of higher quality than lower price point brands, rather than simply saying "Hey, this shampoo is natural, so pay more!"

The term "natural" has not been defined by the US government, and efforts to get them to do so have fallen on deaf ears. Therefore, the term must be understood through a patchwork of international labeling certifications that remain confusing to the average consumer. Although there are some guidelines for "green" marketing in the USA, these are not hard regulations and enforcement will be impossible. The

Food and Drug Administration (FDA) and Federal Trade Commission (FTC) lack the resources and the will to address this issue, so there will continue to be abuse of the term "natural." Certified Organic, on the other hand, is regulated by the United States Department of Agriculture (USDA), and the requirements are much clearer for personal care products than they used to be.

The social trend that has been driving demand for natural, organic, and greener goods and services shows no signs of abating. Every year, concern for health and the natural environment becomes more prevalent. We see it on the news, in our magazines, and in the stores. This general attitude is not a mere fad, but a sustainable trend that has weathered many recessions as well as transcended several generations of Americans. Obesity and environmental problems aren't going away any time soon, so expect industries to address these problems.

Finally, a word should be said about the technological environment. Working with natural ingredients and greener processes can be very challenging. Doing so may require a short-term investment for long-term return, for example. It is simply becoming easier to work with these business inputs every year. Complaints from chemists and engineers, while valid just a few years ago, are largely just that ... complaints.

Much of sustainability involves making products and processes more efficient as well as being safer for people and the natural environment. The Green Imperative provides 12 compelling reasons for organizations to adjust their business models and marketing strategies, and the GRI provides us with a framework for doing so, while ISO-14000 provides us with a global certification for validation. There is simply no excuse, or good reason, to ignore what's going on in the business environment. Business leaders do so only at the peril of the long-term viability of the organization.

REFERENCES

[1] Natural Marketing Institute. www.nmisolution.com.
[2] Ethisphere (2012) Worlds most Ethical Companies. www.ethisphere.com/wme.
[3] Bodylife Cosmetics. http://newbodylife.en.made-in-china.com/.
[4] Pierre-Fabre. http://www.pierre-fabre.com.
[5] Herbal Essences. http://www.herbalessences.com/en-US/hair-care.
[6] Procter & Gamble. http://www.pgbeautygroomingscience.com/home.php.
[7] P&G Reformulating Herbal Essences to Limit Toxins, Greener Design Staff, March 10, 2010, http://www.greenbiz.com/news/2010/03/12/p-g-reformulating-herbal-essences-limit-toxins.
[8] L'Oreal (2011) GRI Data Sheets http://www.sustainabledevelopment.loreal.com/DD/media/pdf/LOrealRDD2011_GRI_Environment.pdf.
[9] Jason. http://www.jason-natural.com/.
[10] Pangea. http://www.pangeaorganics.com/index.
[11] Why Pangea Organics Founder Joshua Onysko Thinks 'Sustainability' is For Slackers, August 2009, Cause Capitalism, http://causecapitalism.com/pangea-organics/.

12

Marketing Case Studies

PART 1: YES TO^{TM} INC.

Ido Leffler and Lance Kalish

12.1.1 THE BACKGROUND

The Yes To^{TM} line of natural products was born out of a genuine desire to break the status quo and create something different in the natural product space. Back in 2005, business partners Ido Leffler (Co-Founder of Yes To^{TM} and Chief Carrot Lover) and Lance Kalish (Co-Founder and Chief Carrot Counter) were looking for ways to lead healthier lifestyles ... from eating healthier foods, exercising more, and choosing "greener" products for themselves and their families. Determined to make a change, the pair sought out natural products to replace their existing skin and hair care regimens. At the time, none of the products on the shelves appealed to the 28 year olds. "We were Mini-driving, Virgin-flying, Apple-using individuals, and we wanted natural products that would appeal to our tastes, emotions, wallets, and style" says Ido Leffler. What they saw in the natural aisle were products targeting a very particular type of "activist" personality, with "holier than thou" stances aimed at scaring the average consumer into using natural products over mainstream alternatives. Lance and Ido immediately saw an opportunity in the market for an affordable, stylish brand that encouraged people to dip their toe into the natural world in a fun, upbeat way. With this in mind, the pair decided to create "The Apple" of the natural beauty world. The fundamentals were simple, summed up in Yes To^{TM}'s "Four Love Points" – the company's mantra (see Figure 12.1.1).

Sustainability: How the Cosmetics Industry is Greening up, First Edition. Edited by Amarjit Sahota.

Figure 12.1.1 The Four Love Points of Yes To™ Inc.
Source: © Yes To™.

12.1.2 THE GROWTH PATH

In 2006, the pair teamed up with a local formulator to create a company that was truly different from any other natural beauty product on the market. They focused on formulas using organic fruits and vegetables, a fun-filled, positive attitude, and a unique brand voice. After only three months and the development of 16 Yes To™ Carrots products, the team got the opportunity to present to US drugstore chain Walgreens. The Walgreens opportunity presented Leffler and Kalish with a very early, company-making dilemma. On the one hand, getting into Walgreens would present a major business coup and allow for the type of production and economies of scale that could ramp up the business quickly. On the other hand, launching in Walgreens, considered a mainstream, mass retailer in the USA, was not something that any other natural beauty brand had done in the past. The industry standard was for natural brands to build credibility and support through a slow, ramp-up style distribution in the nation's fragmented natural health store channel. This was a slow route to market but helped build natural brands a steadily growing and loyal customer support base that could be leveraged when the brand entered the mass channel. At that time, BURT'S BEES® – the leading natural beauty brand – had been around for 21 years and had built a massive and loyal following through the natural health channel over those years. They had then leveraged this base to get distribution into Walgreens, and the gamble had paid off for both BURT'S BEES® and Walgreens. This initial success had actually paved the way for Walgreens' desire to find another natural brand that could emulate the success of BURT'S BEES® and help Walgreens add to this newly-found incremental sales category in their drugstores.

This wave of natural brand success was another factor that led Leffler and Kalish to create the Yes To™ brand, for a new, emerging type of natural product consumer who they referred to as "Light Green" or the "Yes To^{TM} Stylista." This is a term they used to describe consumers like themselves – the evolving consumer who takes steps to lead a healthier lifestyle whether it be via better eating habits, purchasing more organic products, and/or exercising more regularly. These are the consumers who are starting that walk along the path from mainstream products to natural alternatives.

In order to squarely hit this consumer target on the nose and in the biggest way, a partnership with Walgreens would be ideal. So Leffler and Kalish put everything they had into presenting to Walgreens and achieving a partnership and large-scale rollout in the USA. This initial presentation led to an online trial followed by national rollout into 5800 Walgreens stores ... just six months into the life of the business!

With such a massive opportunity laid before them so early in the business's life, Leffler and Kalish experienced their "make or break it" moment in the months between the Walgreens approval for a national rollout, and the launch date slated for five months later. First, they had to find the funding to manufacture and support the production of a multi-million dollar opening order. Second, they had to find a manufacturer and supplier in Israel who was able to manufacture such a large order in the relatively short amount of time they were given. Finally, and most worryingly, they had to finance and come up with a solid marketing plan that would ensure the huge amounts of products going onto the Walgreens shelves would actually be purchased. Otherwise Walgreens had the right to send the products they purchased back and demand repayment.

The task at hand was colossal and not for the faint-hearted. Leffler and Kalish made a lot of sacrifices with their finances, families, and lifestyles. With hindsight, the pair could see the value in building up a natural product company slowly and steadily through independent health stores, giving enough time, experience, and working capital to maximize the chances of success when a mass retail opportunity came along. However, they firmly believed that timing was the most significant element here and there might not be another time in the foreseeable future when a national mass retailer would be open to launching an unknown and foreign brand into national distribution. They were therefore prepared to take up the challenging opportunity.

Using every credit card they had, any personal assets they could find to guarantee loans, and putting the company's equity and their reputations on the line, Leffler and Kalish set out to try and borrow the funds that would allow them to manufacture the inventory required to satisfy the first few orders destined for Walgreens. They partnered with a local manufacturer in Israel and convinced several raw material suppliers to have the faith in them that they were going to build a global consumer brand, and therefore become their largest customers within only a few months. It took a number of "I have a dream" speeches but finally they were successful in getting the support and the essential credit terms that allowed them to produce the first massive order required to pipeline Walgreens. It was the combination of being in Israel (a country used to working on spontaneous and "urgent" projects), and the "never say die" attitude of Leffler and Kalish and their earliest employees that helped push the limits, and enable the production and delivery of a multi-million dollar opening order to Walgreens. The majority of the order had to go air-freight in order to meet the critical timing, however the priority was in meeting the deadline and the order capacity – to overcome that first "make it-or break it" challenge.

Leffler and Kalish negotiated a one-year exclusive contract with Walgreens, which further incentivized Walgreens to try out the brand, and allowed Yes To™ the chance to concentrate its support on one major retailer and develop its resources over the course of the year so it would be better placed to enter other retailers in the future. They knew that a success in Walgreens would lead to countless other retailers in the USA and abroad, so the exclusivity period was mutually beneficial.

Fast forward to 2012 ... Yes To™, Inc. has five core brands distributed in more than 25 000 stores in over 24 countries. The Yes To™ product portfolio has been one of the only natural brands in the world to cross the "retail channel divide" – the products are sold with the same branding in all types of retailers, from drugstores to high-end department stores. Yes To™ is sold by leading retailers, such as Walgreens (USA), Target (USA), Walmart (USA), Krogers (USA), Whole Foods Market (USA), Sephora (Europe), Shoppers Drug Mart (Canada), Waitrose (UK), and Priceline (Australia) covering drug, grocery, beauty, department store, specialty, and mass market retailers around the world.

To get to that happy ending, Leffler and Kalish embarked on a capital raising road show in early 2008 to raise sufficient funds to support the company's growth. In June 2008, the two raised US $14 million from two consumer-focused private equity funds based out of San Francisco. "The key element here was not just to find capital, but to find 'smart' capital" says Kalish, the acting CFO at the time. "What Ido and I needed most were strategic capital partners that could help guide us through the next important steps of growth in the retail world in the US. We decided to go with investors who had already achieved similar success with consumer products in the natural space, and they ended up being a critical component to the brand's continued growth."

The partnership with the new private equity investors was essential and a fundamental driver in the success of the company. Shortly after finalizing the capital raising, Leffler and Kalish moved the main office to San Francisco to get close to the head offices of their new investment partners. This allowed the two new investment firms to play an intimate and integral role in mentoring and assisting Leffler and Kalish in what would be the next critical stage of their growth cycle: moving from exclusivity in Walgreens to a plethora of new, massive retailers. Leffler and Kalish moved themselves and their families to live in San Francisco, and began to build a San Franciscan-based team of sales, marketing, and operations employees, including the employment of a new CEO, Joy Chen. Utilizing the contacts of their new investment partners (who had extensive contacts in the consumer goods industry) they were able to quickly put together an effective team, and start consolidating sales and production in the USA. Over a few years, all manufacturing relocated to the USA; the move helped improve delivery and logistics as well as the cash flow requirements of the company. The company found that growing so quickly in the consumer goods industry can be a two-edged sword; the faster you grow the more cash the company consumes. Consolidating global manufacturing and shortening the cash conversion cycle can immensely benefit young and fast-growing companies, which is exactly what happened with Yes To™ once its production and operations became efficient and consolidated.

Looking back, Leffler and Kalish believe they undertook the large capital raising at the ideal time for the company. Although it was only two years into the company's life, the pair needed the experience of strategic partners who had been there before, and the deep pockets to be able to still survive after a few stumbles during their critical growth period. Not having to continue to go back and seek more funding to support future growth was extremely beneficial in not interrupting the momentum and taking away their management focus, required to take high market share in what was a highly fragmented natural personal care industry.

Yes To™ became the number two natural beauty brand in the US Food, Drug, and Mass channels just four years after its launch, having leapfrogged multiple established brands that had been on the market for decades. Today, Yes To™ only trails BURT'S BEES®, a 27-year-old brand that was acquired by The Clorox Company in 2008 and has dominated the lion's share of the natural beauty market for over a decade, enjoying widespread distribution and strong brand recognition backed by its large corporate owner. However, in recent years, Yes To™ has led the category growth, having grown five times faster than the overall natural personal care category in 2010 and 2011.

12.1.3 MARKETING STRATEGY

With limited marketing funds and an upcoming national rollout in Walgreens just six months into the company's life, Yes To™ experimented with "traditional" advertising, in particular print media. Their first tactic was to engage a media agency and spend a relative fortune for the company on print advertisements in several beauty magazines. The agency employed a "scatter-gun" strategy in an attempt to reach as many print magazines in as little time as possible, primarily by only purchasing distress rate opportunities. Over a million dollars was spent that achieved Yes To™ national brand exposure in a period over six months. The advert that was produced for this campaign, however, lacked any call to action to drive a new consumer to purchase, and was entirely focused on achieving brand awareness. The focus on targeting distress rate opportunities also resulted in more than half of the adverts appearing in the back pages of leading print magazines, making measuring the ROI for this campaign exceptionally difficult. Therefore, on one hand the campaign achieved relative success in getting the new brand's name out there; however, without sufficient brand recognition this failed to move the needle and did not show a strong enough return on investment, measured through the sales seen at Walgreens.

With that learning, and a depleted marketing chest, Yes To™ began to experiment with social media, sampling, and in-store activities to build brand awareness in a more cost-effective way. The return on investment on these tactics proved to be key to the brand's success. In 2008, Yes To™ launched a MySpace campaign to find "The Face of Yes To™ Carrots" targeting US colleges. This viral campaign proved extremely successful, attracting over 150 000 votes and new engagements. Yes To™ continues to this day to invest heavily into social media. Yes To™'s Facebook presence is strong and growing, boasting almost 150 000 fans, and running daily interactions and competitions with thousands of users a day. Yes To™ also has an active social media presence via Twitter, YouTube, Instagram, and Pinterest.

The most successful marketing tactic to accelerate the brand, however, was the in-store programs promoting free trial of Yes To™ products. The offers allowed consumers to try the products for the first time; the offers brought them back to buy more within the brand.

The next major focus was Public Relations. In Ido Leffler, Yes To™ had a strong face and personality (and willingness to wear orange!) to embody the brand. He was able to build relationships with editors, drive editorial interest in the brand, and put a friendly, accessible face with the brand. Partnering with the right public relations firm also contributed immensely to this momentum. Yes To™ was a small brand at the time, and it required a dedicated PR firm that could essentially "adopt" the brand and its unique personality. Rather than find a huge beauty PR firm, Yes To™ sought something much smaller and more intimate, so that the two organizations could work hand-in-hand to promote and nurture the brand in the media world. Yes To™ landed on a boutique firm based in New York. The team had strong relationships in the beauty world and immediately gelled with Ido. Being a smaller firm, they were very nimble and essentially acted as an extension of the brand. This meant they could jump on last-minute opportunities without multiple layers of communication that slowed down huge firms.

As a result of the close collaboration of Yes To™ with the PR firm, the brand has garnered literally billions of PR impressions globally, across every leading magazine, publication, and media outlet (Figure 12.1.2). The firm has also helped Yes To™ achieve a large number of product awards, which are critical to gain industry/consumer credibility in the US market. Worth mentioning is that Yes To™ won two Allure Best of Beauty Awards (in 2011 and 2012), widely considered as the top beauty awards in the USA.

Say yes to Green Say yes to Spas Say yes to Hope
Say yes to Paraben Free Say yes to Health Say yes to Organic
Say yes to Fresh Say yes to Me Say yes to Dead Sea Minerals
Say yes to Curves Say yes to a Better World Say yes to Love
Say yes to Peace Say yes to Shoes Say yes to Rejuvenation

Say yes to

A delicious new way to make your hair shine and your skin glow from head to toe.

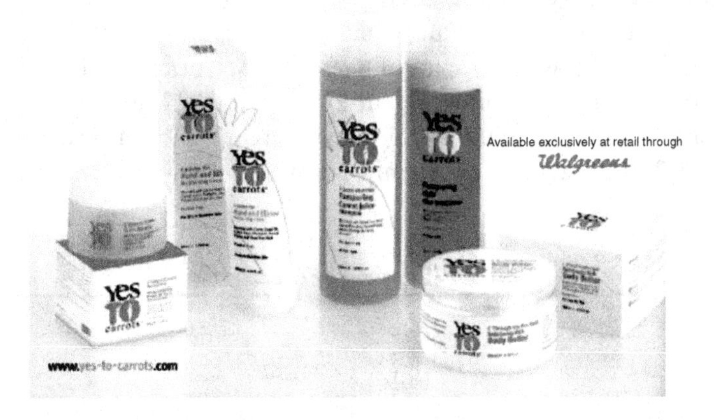

Figure 12.1.2 Print advertisement in the October 2010 edition of Cosmopolitan Magazine.
Source: © Yes To™.

One of Yes To™'s most gratifying and respected marketing efforts has been its efforts and activities in developing the Yes To™ Seed Fund. The Yes To™ Seed Fund is a 501(c) 3 not-for-profit organization that Leffler and Kalish set up around the same time they set up Yes To™ in the USA. The idea was to fund the Seed Fund solely from Yes To™ proceeds and focus on giving back to society. Initially, the majority of the funds donated were to charities or other not-for-profit organizations that were focusing on rehabilitation through gardening and the development of organic farming. In 2012, the Seed Fund is entirely focused on working on projects that promote the building, education, and benefits of growing organic fruit and vegetable gardens for young children. Yes To™ has had its hand in funding the development of dozens of organic vegetable gardens at public schools all around the

USA, and most recently in Africa. In partnership with the US-based not-for-profit organization Mama Hope, the Seed fund finances the creation of micro-farms on school grounds to help supplement meals for African students. The first project in Kenya feeds 1800 kids a day.

12.1.4 PRODUCT POSITIONING

Yes To™'s positioning is one of the major success factors behind its meteoric rise. The brand's focus on the "green stylista" (stylish, fun consumers who are working towards leading healthier lifestyles) helped differentiate Yes To™ from older, more recognized natural brands as well as mainstream non-natural brands. To encourage first-time natural consumers, the brand uses vibrant, fun packaging that explains the benefits of fruits and veggies in a clear, accessible way and strikes a tone that doesn't take itself too seriously. Yes To™ products are offered at an affordable, "guilt-free" price point, which also encourages first-time consumers to dip their toe into the world of natural. Yes To™ nurtured this brand position and undertook a major re-branding in 2010 and 2011 in order to maintain its unique positioning as a nexus between mainstream and natural beauty products (Figures 12.1.3 and 12.1.4). An example of this positioning was verified in 2010

Figure 12.1.3 Brand positioning of Yes To™.
Source: © Yes To™.

Figure 12.1.4 Old branding to new branding of Yes To^{TM} (2010).
Source: © Yes To^{TM}.

by a leading marketing research firm, Organic Monitor, which described Yes To^{TM} as the brand that helped change the US consumer's perception of natural products from *alternative* to *popular*. Yes To^{TM} disrupted natural product industry norms by proving you can launch first in mass market retailers (as opposed to natural health stores), and still have the credibility to win consumer minds and acceptance.

12.1.5 DISTRIBUTION GROWTH AND BRAND EXTENSIONS

After an exclusive period of 12 months with Walgreens, concluding in mid-2008, Yes To^{TM} began to roll out to other major retailers in the USA, such as Target, Duane Reade, and Safeway. The decision was made to target all channels (food, drug, and mass) and develop a strategic retailer within each channel. By 2011, Yes To^{TM} was selling in over 30 major retailers in the USA, covering nearly 22 000 stores across the country (Figure 12.1.5). Walgreens remained the number one drugstore, Target the number one mass market retailer and number two customer overall, and Safeway became the number one grocery retail partner.

Internationally, Yes To^{TM} launched in 12 countries within its first year of operation. Initially the pace was required to help fund the US business and tremendous growth of the company. However after funds were raised in mid-2008, a more strategic and direct approach to international markets began to unfold. In April 2009, Yes To^{TM} struck an exclusive distribution deal with Sephora Europe, which agreed to carry the Yes To^{TM} brand in all stores throughout its network of 500 stores in 12 European countries. This mid-2009 rollout has been extremely successful and proved that Yes To^{TM} could be sold in high-end European perfumeries, as well as in the mass market in the USA, with the exact same branding and packaging. Yes To^{TM} has opened up additional European markets in the Netherlands,

Figure 12.1.5 Yes To™ Products in Walmart – anchoring the natural beauty category.
Source: © Yes To™.

Belgium, Germany, Russia, the UK, Ireland, and Denmark, primarily appointing distributors in those countries that are still in place (2012).

In 2011, Yes To™ started to take distribution in some country markets under its own wing. This has proved to be a very successful strategy, especially in Australia. In 2012, Yes To™ Inc. is in 24 countries and is expanding rapidly through a mixture of distributor agreements and direct management.

12.1.6 FUTURE PLANS

Even after its unprecedented success in distribution and brand awareness since its launch five years ago, Yes To™ sees significant growth opportunities within traditional Food Drug and Mass retail channels (FDMC) over the next few years. Yes To™ plans to expand its share of the natural beauty category. Initially launched exclusively in Walgreens for a year, followed by a strong partnership with Target, Yes To™ has a particularly strong market presence in the Drug and Mass trade; it plans to consolidate its position as the clear number two natural player in the

US FDMC. The natural beauty category in retailers has increased considerably over the past 10 years, assisting the growth of Yes To^{TM} in retailers that have been slower to adopt natural products. With Yes To^{TM}'s unique positioning and position as a leading natural beauty brand, it has greatly benefited from this organic market growth.

The focus of Yes To^{TM} for the last five years has been clearly on the US market, leaving potential for massive growth in international markets. To date, Yes To^{TM} has enjoyed global expansion with major retailers in some of the world's strongest beauty markets, primarily using third-party partners with management from San Francisco. Moving forward, the company will explore putting greater resources in some of its major distribution markets, with Australia a priority. Asia is a major component of Yes To^{TM}'s strategic plan for growth; exciting partnerships with retailers are scheduled to be announced in the next two years.

Major advantages of the Yes To^{TM} brand are its simplicity and its universality. The brand was deliberately designed to enable its expansion into other consumer categories. Yes To^{TM} has a lengthy pipeline of new product development that continues to push the limits of developing better, more effective, and natural alternatives to popular mainstream items, which has become a strategic point of differentiation for the brand. Yes To^{TM} will also continue to expand its range by launching exciting new Yes To^{TM} families under different fruits and vegetables.

PART 2: KORRES NATURAL PRODUCTS

George Korres

12.2.1 HERBAL REMEDIES TO NATURAL PRODUCTS

Korres is a Greek skincare brand, born out of Athens' first ever homeopathic pharmacy in 1996. The first Korres product was an aromatic herbal throat-soothing syrup with honey and aniseed; a recipe inspired by "*rakomelon*," a warming spirit-with-honey concoction, which the grandfather of the company founder George Korres used to make in his hometown on the island of Naxos.

12.2.1.1 The Pharmacy

The Korres story begins at the oldest homeopathic pharmacy of Athens, located in the city centre near the Kallimarmaron Stadium. George Korres joined the pharmacy team in 1989 while still studying Pharmacology at the University of Athens. Though skeptical at first about homeopathy, George was soon overwhelmed by the power of natural ingredients. Driven by his own passion and deep understanding

Figure 12.2.1 George Korres at the Pharmacy.
Source: Korres Natural Products.

of herbs, George envisaged the development of a natural skincare line, which was realized soon after he was entrusted with the management of the pharmacy, in 1992 (Figure 12.2.1). His first step was to set up an exemplary homeopathic laboratory based on international lab standards. Led by the pharmacy heritage, his knowledge and understanding of over 3000 herbal remedies, and his quest for a more natural approach, he started developing formulations that were first tested on his own friends.

In 1996, Korres was born. The brand's simple philosophy was rooted in four core principles that remain the same as in its early pharmacy days: natural and/or organic ingredients; skin and environmentally friendly products; clinically tested efficacy; affordable and interesting esthetics.

The first product was the Korres Wild Rose 24-Hour Cream, an early innovation and a global best-seller ever since. The pharmacy still remains the city's biggest pharmacy of its kind with a dedicated clientele while also serving as a training school for new Korres staff members.

Homeopathy was the starting point; inspired by its "mild but effective approach" yet focusing solely on nature as the ultimate provider. Natural ingredients are selected because of their properties, are extracted, isolated, stabilized, and tested exhaustively so as to further access their action in relation to skin needs. The homeopathic heritage provides the data, the lab provides the clinical efficacy, the formulations provide the natural alternative to conventional skin solutions.

12.2.2 THE CHALLENGE

Not sharing other countries' advanced cosmetics industry status makes Greece inevitably a challenging environment to support a business idea, especially long before the "green beauty" concept was even deemed fashionable. Considering that at the time the majority of ingredients did not exist in organic form and that Greece did not have the industry to process its endemic plants for cosmetic applications, launching the business was extremely high risk. Still, Greece presented a great opportunity; its rich native flora and the country's endemic plants that exceed 1200 species.

Identifying the endemic flora as a unique opportunity, Korres had to overcome two major sets of difficulties; one associated with the country's cosmetics industry and one specific to the natural cosmetics industry. The company focused on:

- educating on the benefits of a more natural way of living/consuming;
- helping farmers certify their produce for cosmetic applications;
- developing a first-for-Greece herb extraction unit;
- establishing a network of research partners focusing solely on skin; biochemistry related to natural ingredients.

12.2.2.1 Today

The brand's four core principles – as outlined earlier – remain the same since the initial pharmacy days; despite not having an international customer in mind when Korres first started, these core principles have universal appeal and helped establish Korres as a global brand.

Korres grew into a respected brand by focusing its research and development efforts on addressing skin care needs with natural ingredients without compromising the benefits associated with conventional skin care, and developing natural products to help solve conditions that traditionally required heavy chemicals.

Over its 15-year history, the Greek skin care brand counts three worldwide innovations; the first natural 24-hour moisturizing face cream and a global best-seller (Wild Rose, 1996); the first cosmetic formula in the world to incorporate edible yogurt (Yoghurt Cooling Gel, 2003); the first cosmetic worldwide based on the Nobel-awarded discovery of the Proteasome – a result of primary research on "Proteasome and Ageing" – the first natural cream to achieve higher scores on all comparative crash-tests with the market's top-selling anti-aging creams (Quercetin and Oak, 2009).

The Korres portfolio now includes over 400 natural and certified organic products, with brand presence in 30 countries. In addition to stand-alone Korres stores – 25 stores in cities such as Athens, New York, Paris, Madrid, and Singapore – products are also in exclusive department stores in Tokyo, Los Angeles, Milan, Berlin, Sydney, Hong Kong, and in over 6500 pharmacies in its homeland, Greece.

From one single pharmacy, one product, and a team of two, Korres has grown to an over €40 million business (the company ended 2010 with approximately €44.1 million sales). The company was floated on the Athens Stock Exchange in March 2007, in what was described by financial circles as the "*most successful stock market company debut in years.*"

Korres was awarded the World's Most Innovative Brand by International Cosmetic News (2008); the company first entered the UKs CoolBrands® list in 2006 and has stayed there since. The same year – and every year since – it was ranked amongst Europe's 500 highest developing companies (Entrepreneurs for Growth association, Europe's 500). In its 15-year history, the brand has introduced two patent-pending technologies. Since 2010, Korres, along with eight other partners, has been undertaking one of the greatest research projects in the field of natural ingredients: AGROCOS is studying over 3600 natural actives with a view to sourcing nature's most "powerful" ingredients (From Biodiversity to Chemodiversity: Novel Plant Produced Compounds with Agrochemical and Cosmetic interest).

12.2.3 VISION AND STRATEGY

The brand's statement is best encapsulated in the disclaimer accompanying its logo: Herbs, Cooperations, Extractions, Research. Its vision includes:

- HERBS: developing innovative and clinically efficacious skincare products based on natural active extracts.
- COOPERATIONS: educating, supporting, cooperating with local communities, agricultural unions, organic farmers on cultivating and harvesting.
- EXTRACTIONS: utilizing valuable Greek herbs through the production of brand-own, premium organic extracts.
- RESEARCH: primary and applied research on natural ingredients in relation to skin biochemistry.

12.2.3.1 Herbs

All Korres formulations are based on natural ingredients. The Greek flora is renowned for its wealth as it numbers over 6500 varieties, 1200 out of which are endemic plants that are not found elsewhere in the world. Korres, in association with the Agricultural University of Athens, has mapped out the native flora based on climate and soil criteria. The Korres plant hunters select the ideal cultivation areas for each herb based on this mapping while at the same time working towards protecting herbs threatened with extinction.

12.2.3.2 Cooperations

Having selected the ideal cultivation area for each herb, Korres then collaborates with that area's micro-farmers, agricultural unions, and local communities; the company's agriculturists teach them organic farming and sustainable agricultural systems, and help them throughout all stages of cultivation and collection in association with the Agricultural University of Athens: Korres supports them financially by purchasing their produce to be used in the formulations. The choice not to cultivate its own herbs is a conscious one. The company's aim is not to compete with, but to support the farmers and their unions, as well as the community bodies.

12.2.3.3 Extractions

Working closely with the Pharmacognosy School of the University of Athens, Korres has created an exemplary environment friendly herb extractions unit. This first-for-Greece unit enables Korres to extract and isolate active ingredients to produce superior quality, clinically effective, organic/natural extracts.

12.2.3.4 Research

The Korres Research & Development Lab works with an international independent scientific board. This board consists of high-profile professors and researchers who explore the clinical benefits of natural active ingredients in relation to biochemistry, pharmacognosy, and clinical dermatology. Based on primary and applied research, as well as its ongoing commitment to deliver the most natural formulations available, the lab develops products at the company's certified production facility (Ecocert/ISO 9001/ISO 14001).

The use of specific synthetic compounds – such as petroleum derivatives, silicones, propylene glycol, and parabens amongst others – is avoided and replaced by effective natural ingredients in product formulations. Through an analytical, transparent table – Formula Facts – featured on its product packaging, Korres makes extended reference to the formula in an easy to decode and comprehendible way, clearly stating the natural content of each product.

12.2.4 TARGET AUDIENCE

In terms of product development, Korres has made the decision to evolve based on instinct instead of being driven by market research/consumer audits; customer interaction is encouraged but not through guided focus groups.

The brand's core audience as profiled through brand intelligence data – data from the brand's points of sale in Greece, including five stand-alone Korres stores in Athens and the stores abroad – consists of informed individuals who seek natural choices, performance, and inspiration.

Being affordable for everyday use – one of the brand's core values since the beginning – means that spending power is not imperative; Korres consumers are best understood through their choices; people who believe in the power of natural ingredients or wish to be introduced to a "greener" lifestyle; people who care about not only about what goes into a product but also what goes into making it; informed individuals who want to know all the facts and make their decisions based on transparent communication; people who seek artistry in all their choices and have a great understanding of/respect for design and aesthetics.

12.2.5 THE PORTFOLIO AT A GLANCE

All products – natural and/or organic certified – are developed based on four groups of natural ingredients: plants with medicinal properties (Korres pharmacy heritage), endemic plants (Greek Flora), food ingredients, and high-efficacy natural actives.

The brand's portfolio can be best understood through its zones:

1. SCIENCE: Skin care, Hair care, Sun care.
2. COMFORT: Body care, Oral care, Men.
3. LUXURY: Fragrance/Make Up.
4. PHARMACY HERITAGE: Pharmaceuticals (Syrup, Pastilles), Organic Tea Blends, Food Supplements, Herbal Compounds, Essential Oils.

In Greece, Korres is marketed as a dermocosmetic brand and its products are available in pharmacies. The key product category is skin care, with face-care and anti-aging products being most popular. The Quercetin & Oak (Q&O) line is currently the best-seller, outperforming Wild Rose which has been the best-selling item for 15 years. In 2011, Q&O product sales increased by 58% despite stable pharmacy sales in anti-aging dermocosmetics. The Q&O collection comprises up to 45% of the brand's total US sales: the US market is the biggest market for Korres products outside Greece.

12.2.6 MARKETING AND POSITIONING

12.2.6.1 A Leading Natural Player that Talks Efficacy

One of the most frequently asked questions to those in the natural skin care industry is "yes it is natural but does it work?" The popular belief for years was that natural creams can only cater for basic moisturizing needs.

Anti-aging is the ultimate beauty-world challenge and a segment in which visible benefits and high-score results are of extreme significance. Even though the Korres philosophy has been built on no-overpromising, the brand made a tremendous effort to highlight that natural/organic-based cosmetics can now "speak" the language of conventional skincare and debunk the theory that "*only conventional age-defying creams work.*"

Korres's patent-pending technology in its Quercetin & Oak line was a milestone not only for the company but also for the cosmetics industry. The line was based on a Nobel awarded discovery (Proteasome and Aging), and the result of five years of primary research; it was the first natural skin care proven to outperform conventional "supercreams."

The Quercetin & Oak line has undergone independent tests including blind-trials and crash-testing against the market's top performing anti-aging products, scoring an extremely high number on various attributes like wrinkle decrease. Hailed as one of the World's Best Products (*Elle*, US Edition, 2010) and one of the Top 25 Miracle Products That Can Change Your Life (*Marie Claire*, US Edition, 2010), the Quercetin and Oak line has attracted the interest of the scientific community through exploring how to boost naturally a biological path that is the center of scientific interest both in medicine and cosmetology.

The Q&O line was crash-tested against leading products in each category, outperforming them in all aging biomarkers:

- a dermocosmetic based on Retinol;
- a selective-market/ luxury anti-aging cream based on peptides [€150–200];
- a US dermatologist brand [$180].

Furthermore, *in vitro* studies on fibroblasts (epidermal cells) showed Oak derived Quercetin's impact on vital biological aging markers:

- up to 39% reduction of free radicals within cells;
- repairing activity on already aged cells which revert to the young cell phenotype (photographed under the microscope);
- stimulating cell growth and protein synthesis within 24 hours.

In vivo clinical results*

- 85% of users reported visible wrinkle reduction;
- 100% of users reported significant increase in skin elasticity;
- 100% of users reported significant skin texture improvement (softer and more even, compact skin).

*Self-evaluation test/4 weeks of use/dermatologist supervised.

12.2.7 BEAUTY MADE HONEST

Clinical efficacy aside, the second greatest issue of the natural skin care industry is certification. Amid the overload of information and confusion surrounding natural ingredient-based products, and with the lack of a global certification system, Korres launched its Formula Facts initiative.

> Sensing an urgency in the skin care market for labels that are transparent about what is in or is not in a product, Korres Natural Products is taking action now
>
> reported *Women's Wear Daily* back in 2008.

Korres was the first cosmetic brand to decode the formula ingredients using everyday language, addressing thus the formula synthesis/content in such transparent and honest way.

Sincere and transparent communication with those selecting Korres products has always been of extreme significance to the company. The practice of "greenwashing" has been extensively and growingly employed in cosmetics. Many brands make all-natural claims despite only using an extremely low percentage of natural ingredients, and/or stress that they avoid specific substances when instead they are using several others that can be even more harmful. Having realized the overload of information and the level of misunderstanding surrounding natural ingredient based cosmetics, Korres introduced a transparent and analytical ingredient table on its packaging that makes extended reference to the formula itself and its natural content, in a way that everybody can comprehend, while also addressing frequently asked questions such as the ones about animal testing or vegetarian/vegan-friendly use.

The Korres philosophy has been built on natural formulations but with safety and efficacy on top of the agenda; it made the decision to stand by safety, efficacy, and natural direction in this very order instead of sacrificing its core principles to be able to make 100% natural claims. The Formula Fact initiative enabled Korres to inform/educate consumers and explain why specific synthetics were avoided or selected.

12.2.8 SUSTAINABILITY

The Korres sustainability effort is multi-faceted and drives every decision made.

12.2.8.1 Ingredients

By protecting Greece's 1200 indigenous flora, Korres protects the world's flora. All formulations are based on natural or certified organic ingredients. The Korres plant hunters select the ideal cultivation region for each herb while at the same time working towards protecting endangered herbs and sustaining plant populations. Korres avoids the use of harsh chemicals and substances that can be harmful not only to the skin but also the environment. The use of non-biodegradable ingredients is eliminated.

12.2.8.2 Energy

Korres follows a voluntary carbon footprint monitoring program based on a credible, simple set of criteria that represent real, quantifiable emission reductions. Korres' company headquarters – including production, packaging, storage and offices – operate entirely on 100% renewable energy. The Green Guarantee certified plant based on the outskirts of Athens is also ISO (14001: 2004) and Ecocert certified.

12.2.8.3 No Waste

The plant features an environmentally friendly, first-for-Greece plant extraction unit that uses natural instead of chemical solvents in spite of higher cost. By operating a no waste policy, Korres processes the raw ingredients in a manner such that all parts of the plant are used, enabling the production of up to 11 derivatives from every single plant. The residue from the process is turned into organic soil and sent to the company's network of organic farmers.

12.2.8.4 Responsible Packaging

Almost all of Korres packaging is recyclable, and when possible recycled. For its packaging, the brand uses environmentally friendly materials; each product is designed so that its individual packaging parts can be easily separated and recycled. Unnecessary packaging is eliminated. The entire Korres Hair & Body portfolio is box-free. Packaging facts are featured on the Formula Facts table of each product.

In-house, there is a thorough recycling policy for printing, plastic, paper, glass, electric, electronic, retail and merchandise units.

12.2.8.5 Community

12.2.8.5.1 Organic Farming

Korres collaborates with micro-farmers, agricultural unions, educational and social institutions. The in-house agriculturists teach the company's partners organic farming and sustainable agricultural systems helping them throughout all stages of cultivation and harvesting. Korres supports its partners financially by purchasing their product. In addition to the cultivation standards that form the basis of these partnerships – selecting the best cultivation area for each plant – Korres also takes into consideration social criteria too, for example supporting rural areas with high unemployment.

This continuous effort is part of an overall focus on ethical cooperation that includes a network of suppliers consisting of people with special needs. This network caters for a percentage of the company's promotional materials including printing and sewing productions.

12.2.8.5.2 Protecting natural habitat

The Friends of Oak & Nature union mission is to protect the biggest Oak forest in southeast Europe. Located in northwestern Greece, the 130 000 sq m forest hosts:

- 256 endemic plants
- 126 bird species, 26 out of which are threatened with extinction
- thousands of great-age oak trees

The forest is mainly threatened by illegal wood-cutting and arson. Korres contributes towards the replacement of illegally cut trees with an annual fun while cooperating with the local authorities on an effective protection plan.

12.2.9 GLOBAL PRESENCE

Despite the fact George Korres did not set off with global intentions, within three years of setting up Korres, the first export deal arrived. The brand's global journey started accidentally from a pharmacy on the island of Crete in 1999. An American buyer spotted the products and asked to be the first to import them to the USA. In 2000, a beauty buyer from Harvey Nichols saw the products in New York and placed an order for the store's newly created *Beyond Beauty* section which featured niche brands from around the world. This move initiated a number of partnerships with prestigious department stores in Britain and France, including Selfridges, Harrods, Liberty's, Bon Marche, and Galleries Lafayette. Exporting was not a strategically planned move, however it resulted in a domino-style effect, allowing Korres products to enter markets in many parts of the world.

The first stand-alone Korres store opened in London in 2002 at a time when brand awareness outside Greece was zero. A bold move, yet one that played a significant part in establishing Korres as an international brand. The brand's strategy in those early days was to achieve desirability status through its presence in the most prestigious stores worldwide; once that was achieved, the second step was to build awareness through communicating its unique story. This was a twofold effort requiring media exposure, along with a platform to showcase its entire product portfolio (vs. the limited shelf space in the department stores) and communicate its key messages. The stand-alone store in one of the world's best destination cities was a vehicle to allow faster international exposure and boost awareness.

Two more stand-alone stores opened in Barcelona (2004) and then the US breakthrough followed thanks to the Sephora network that recognized the serious product development effort of the pharmacy-born company; what was considered a small, Greek, cult brand started winning over the audience of a mature and hugely demanding market; a market that was driven by high-tech, "miracle" products with overpromised claims.

With the continuous media, industry, and consumer support, Korres started gaining equity and built a portfolio of over 400 products with a worldwide presence that exceeds 5500 selective and semi-selective distribution points, among which are high profile concept shops, exclusive department stores and pharmacy chains, in addition to the brand's online selling points, its shop-in-shop points, and stand-alone stores.

Secondary to its growth, yet as important to its brand awareness, are the brand's duty free and on-board presence (the brand in its 15-year history has had had a number of business class amenity deals, including Lufthansa, Aeromexico, Tam, Etihad, and Delta airlines) as well its network of exclusive hotels around the world, including Morgans in New York, Delano in Miami, St Martins in London, and Hilton in Tokyo.

Korres is currently developing its diversified expansion abroad, moving from the initial "global awareness" phase to a deeper mature-market penetration strategy. The brand's natural retail environment, due to its roots, herb heritage, and research focus, is the pharmacy distribution network for markets where natural cosmetics are predominantly in pharmacies.

Due to the brand's equity – natural and clinically efficacious, ethical, pleasing to use, affordable, and of interesting design – Korres sits equally well in high-end, luxury department stores as specialized cosmetics retailers and independent beauty/perfumery shops. In department stores, products are usually in the natural cosmetics section or in the para-pharmacy area where available. In some cases, the brand has managed to secure shop-in-shop areas within the cosmetics hall.

12.2.10 A CLOSER LOOK

12.2.10.1 Europe

Korres has a presence in most markets, focusing on those with strong cosmetics background (France, Spain, Italy, the UK); those that are more receptive to "green" cosmetics (Germany, Scandinavian market) and those that are smaller yet upcoming (Cyprus, Poland, and the Czech Republic).

Its strategy is strengthening its presence in the so-called mature markets (Germany, Spain) while also investing in those that present great potential (Russia, the UK, Belgium, Scandinavia). The expansion model for each market varies. For instance, Spain is very similar to the Greek market, with the pharmacy network the major driver for growth; Korres products are sold in over 1100 pharmacies, 35 department stores (El Cortes Ingles), and four stand-alone stores. In Russia, where the pharmacy concept is not as strong, Korres is cooperating with the two biggest cosmetic store chains that cover 73% of the market.

12.2.10.2 USA

Similarly, in the USA, where the apothecary notion may be appealing but there is no traditional pharmacy environment, the brand has focused on the Sephora network, ULTA, SiJCP, and a number of selective and prestige points throughout the country, for example, Fred Segal in Los Angeles. It has also explored the possibility of increasing its volume through alternative channels including HSN television shopping channel and online retailers, for example, the brand's own e-shop (www.korresusa.com). Korres entered an agreement with Johnson & Johnson US to partner on product development, manufacturing, sales, and distribution in North and Latin America (effective as of 2011).

12.2.10.3 Asia and Australia

In Asia (China, Singapore, Japan, and the United Arab Emirates) Korres products are in selective high-end department stores and cosmetic retailers that include Harvey Nichols in UAE, Sephora in Singapore, and Isetan in Japan. Korres also has stand-alone stores in Singapore and UAE (Figure 12.2.2). In Australia, Korres is in partnership with Kit Cosmetics, a chain of beauty stores specializing in niche skin, hair and body care products.

Figure 12.2.2 Korres Store in Dubai.
Source: Korres Natural Products.

12.2.11 A SUCCESS CASE STUDY STARRING ... THE PRODUCT

The brand's focus is on raising awareness, enticing new and existing customers while most importantly communicating its philosophy, values, and core messages; transparency has always been key in the process, long before it became a trend for brands to showcase ethical production techniques or share exactly how and where they make their products. The brand's origin also plays an integral part in its growth. The Greek heritage has helped introduce the product, yet it is the product itself, its performance, and what it stands for, that has established Korres. Marketing support is integral – each market develops a localized marketing plan based on the brand guidelines and executed through a combination of basic and creative marketing tools – but the starting point is the product. The ongoing effort is to safeguard the brand's image and ensure that this is communicated in a consistent way across the globe.

12.2.12 THE FUTURE

Research will remain fundamental to the company's growth in terms of product development; identifying natural ways to activate specific biological paths that enable cells to prolong their healthy life-cycle. Korres realizes that science and technology today have made possible the use of several raw materials in a way that allows development of clinically effective, natural, and sustainable formulations that are as good as conventional solutions yet more skin and environment compatible, without compromising on performance.

To that end, the company has just strengthened its anti-aging and hair care portfolio (2012 launches) consistent with its continuous effort to introduce natural problem-solving products addressing key concerns that up to now required heavy chemicals. The year 2014 will see the completion of the AGROGOS research program, the findings of which are expected to be a milestone for the natural beauty industry.

Studying the cosmetic properties of 3600 natural actives, focusing on their anti-aging, anti-oxidant, sun-protecting action (anti-aging, anti-hyper-pigmentation and UV-protection), the research will discover, and carry to the stage of development candidates, plant-derived small molecules with potential as new cosmetic agents. These compounds will derive from plants originating from major biodiversity hotspots in Europe, Africa, Latin America, and the Asia-Pacific region.

In terms of its global presence, Korres is set to continue growing, through targeted initiatives focusing on its top international markets, aiming to further strengthen the brand's presence in both mature markets (the USA, the UK, Germany, Spain) and those that present great potential (Russia, Scandinavian countries, Latin America).

PART 3: WHOLE FOODS MARKET

Jody Villecco

12.3.1 INTRODUCTION

Founded in 1980 in Austin, Texas, Whole Foods Market® is the world's leading natural and organic food retailer. Whole Foods Market's motto, "Whole Foods, Whole People, Whole Planet™" captures its mission to ensure customer satisfaction and health, team member excellence and happiness, enhanced shareholder value, community support, and environmental improvement. Thanks to the company's more than 65 000 team members, Whole Foods Market has been ranked as one of the "100 Best Companies to Work For" in America by FORTUNE magazine for 15 consecutive years since 1998. In fiscal year 2011, the company had sales of more than $10 billion and currently has more than 315 stores in the United States, Canada, and the United Kingdom.

Whole Foods Market's culture and the success of the company is driven by its strong core values, including "selling the highest quality natural and organic products available" which is rooted in the company's stringent quality standards.

Crafted to help its buyers seek out the highest quality natural products that support health and well-being, these strict quality standards provide a touchstone that gives substance and integrity to the brand and are the basis for the company's customer and team member trust. In essence, Whole Foods Market does the "homework" for their shoppers.

With "no artificial colors, no artificial flavors, and no artificial preservatives" as one of the company's founding principles, further refinement of Whole Foods Market's quality standards began in the later 1980s, largely due to the beginning of expansion into new markets. At the same time, the food industry, in general, was increasingly introducing new ingredients and new products, further accentuating the need for a more definitive guide for the company's buyers to ensure consistency when identifying acceptable products within its stores.

12.3.2 BODY CARE QUALITY STANDARDS

Whole Foods Market's commitment to upholding the highest standards extends far beyond the food products it sells, with equal commitment to providing superior options and transparency for its personal care products. Accordingly, as the company's selection of body care products slowly grew, so did the need for creating similar more definitive guidelines and standards for non-food products. Whole Foods Market's original Body Care quality standards list reflected an early view of natural personal care: no artificial colors in body care products, and a select number of unacceptable preservatives and other ingredients. While simplified, these original body care standards were more stringent than what was required in conventional markets; they had already set Whole Foods Market apart as a leader in the natural personal care products industry.

Whole Foods Market's attention to Body Care quality standards intensified shortly before 2008, the year that the company's top tier Premium Body Care quality standard was launched. Over the previous years, the company had received an outpouring of feedback from inquisitive shoppers who looked to Whole Foods Market to help them pick the cleanest, most natural personal care products. Given the expanding selection of natural personal care products on the shelves at Whole Foods Market, along with unfamiliarity with the nomenclature of certain personal care ingredients, customers reported they did not know how to identify the most natural products available, nor did they know how to distinguish between natural and conventional products. They also wondered why certain ingredients were found in products on Whole Foods Market's shelves, as they expected the grocer to carry the most natural and highest quality products, ones distinctly different from conventional venues. At that time, there were no other reputable natural personal care standards to lean on for guidance. So with the intent of providing clear guidance for its customers, Whole Foods Market embarked on enhancing its Body Care quality standards.

With Premium Body Care, Whole Foods Market made the decision to develop a top-tier quality standard, as opposed to reformulating a stricter set of entry-level standards. At its core, the decision originated from a spirit of partnership with its vendors. The company wanted to allow manufacturers ample time to reformulate their products, in order to inspire innovation, as well as to give manufacturers a marketplace incentive for improving their formulations. It would also offer customers a spectrum of choices when it came to the kinds of personal care products that were available.

In the process of drafting its top-tier Premium Body Care quality standards, Whole Foods Market took an objective approach, enlisting the help of cosmetic chemists with experience in both natural and conventional cosmetic formulation. It also utilized its own technical staff and the technical assistance offered by vendors and ingredient suppliers. To start, a survey of nearly every ingredient found in products on their store shelves was gathered and divided into functional classes. At this point, the ingredients were evaluated one by one and screened, not only for their source (or "naturalness") but as was done for foods: the safety, environmental impact, and the efficacy of the ingredients were all examined. The overarching questions the quality standards team applied to all ingredient categories included: what are the mildest, safest alternatives and most natural ingredients available? And, which ones are absolutely necessary for the product to work well?

It was a challenge for some personal care ingredients to pass all four Premium Body Care criteria, and sometimes compromises had to be made depending on priority. First and foremost, all ingredients had to be safe, with no indication of harm for the general population. Where possible, existing scientific and technical data were utilized to evaluate each ingredient's safety; but in the absence of hard data, decisions were based upon the precautionary principle, especially if there were safer functional alternatives known to be available. Due to safety concerns, formaldehyde-releasing preservatives, chemical sunscreens, and similar chemicals were automatically eliminated. Synthetic fragrances were also excluded as they can be highly irritating to sensitive individuals; they are also a potential source of questionable sub-ingredients. Other decisions bordered on the philosophical, with the necessity to make judgment calls based on priorities. For example, certain ingredients, such as chelators, that do not have any associated safety concerns were disallowed from Premium Body Care because they had a negative environmental impact. Conversely, ingredients that were not 100% natural – such as gentle surfactants – were occasionally allowed due to their functionality. A key goal was for all Premium Body Care products to function similarly to their conventional counterparts. In the end, the Premium Body Care standard is unique to Whole Foods Market; every ingredient has a rationale for its status, with the unacceptable ingredients listed on the Whole Foods Market public web site [1] in the spirit of transparency. With many hours spent reviewing ingredients and products, this proved to be a fabulous learning process for all parties involved.

Manufacturers were invited to apply for Premium Body Care designation at no cost, except for the time needed to fill out required paperwork. Upon expressing interest in obtaining Premium status, each manufacturer was given a vendor packet, which explained the Premium Body Care review process and included all necessary forms. For each product, the manufacturer was required to disclose all of the ingredients in the formula, as well as to submit product labels. The Premium Body Care review was conducted by members of the Whole Foods Market quality standards team. The process involved ingredient screening, an extensive label review for compliant claims and proper ingredient listing in International Nomenclature for Cosmetic Ingredients (INCI) if applicable, examination of paraben or particle size test results, and a thorough fragrance review. Fragrances were one of the most challenging areas to research and review for Premium Body Care because there has been no official or regulatory definition in place for natural fragrances. Furthermore, fragrance houses are notorious for being secretive about their fragrance compositions. Since the testing and verification of fragrances is complex, it is nearly impossible to determine if a fragrance is 100% natural or not. For this reason, a fragrance questionnaire was developed by Whole Foods Market as a required component for application to Premium Body Care. Products that passed all criteria were added to the master Premium Body Care product list; the standard was launched with about 1200 acceptable products. In stores of some regions, customers were able to identify which products qualified for Premium Body Care on store shelves via Premium Body Care logo shelf-tags. In 2012, store-specific Premium Body Care lists became available on each store's web site via the public Whole Foods Market web site [2].

Other educational tools were made available to both customers and Whole Foods Market team members. Foremost was a designated Premium Body Care area within the company's web site [3] that presented a brief, customer-friendly explanation of the standard along with a link to all unacceptable ingredients. Premium Body Care brochures [4] have been available on an ongoing basis in stores and on the company's web site. Whole Foods Market has also posted an online educational Premium Body Care video [5] together with supporting podcasts. Premium Body Care has been periodically covered in the Whole Story blog, with several notable posts. One highlighted the importance of the vendor connection to the success of Premium Body Care; another provided an in-depth distinction between Premium Body Care products, entry level body care products, and conventional products. Another educated consumers about sunscreens, explaining which ones are Premium [6–8]. Moreover, team member education has been critical to the promotion of the program, such that the company has invested in online and in-person training, as well as sustained communication about Premium Body Care with the teams. In addition, the company's team members have access to Premium Body Care

Figure 12.3.1 Whole Foods Market Premium Body Care Stands in Stores.
Source: Courtesy of Whole Foods Market. "Whole Foods Market" is a registered trademark of Whole Foods Market IP, L.P.

reference "apron cards" for use while assisting customers in the departments. Vendors have also supported and played a role in these educational initiatives, reinforcing the program during their in-store training and actively touting the standard and Premium Body Care-approved products in their marketing materials. Finally, the program has received extensive and favorable coverage in the media, further underscoring its impact on the natural personal care industry. (Figures 12.3.1 and 12.3.2.)

In late 2012, the Whole Foods Market's Premium Body Care program had been in place for more than four years; it currently has about 4000 acceptable products. Premium Body Care-designated products comprise a far greater percentage of the personal care products on the company's shelves than they did at the program's launch, achieving one of the original goals of the program. The increase is due to the Whole Foods Market purchasing team's strong support of the program, and their commitment to work with suppliers to inspire the development of products that meet the standard. For example, Premium Body Care products have been preferentially accepted as new products, whilst reformulation to meet the higher standard has been actively encouraged, especially with core vendors. Many of the company's own 365 Everyday Value personal care products were reformulated in response to Premium Body Care, providing more affordable Premium options for customers.

However, it is also important to note that it does not need to be all or nothing when it comes to moving the bar forward to "cleaner" personal care products. For

Figure 12.3.2 Premium Body Care Banners for Whole Earth Campaign.
Source: Courtesy of Whole Foods Market. "Whole Foods Market" is a registered trademark of Whole Foods Market IP, L.P.

example, even among products that do not meet the Premium Body Care standards, many have been reformulated to remove non-Premium ingredients; the net result is that the quality of the entire personal care department at Whole Foods Market has collectively improved during this time. These wide-scale reformulations can be partially attributed to the proven track record of natural personal care ingredients; there is no longer a step down or trade-off in functionality of such products compared to conventional personal care products. With constant innovation and scientific research supporting the benefits of many new plant-based ingredients, natural personal care products are increasingly perceived by customers as on par with mainstream products and efficacious in their own right.

It should also be acknowledged that this change has been made possible by the greater availability of safer, more natural, raw materials because of the dedication of innovative raw ingredient suppliers, as well as the willingness and enthusiasm of vendors to reformulate unacceptable products. Changing the formulation of a personal care product can be challenging, often requiring time, expertise, and persistence to be done properly. When a personal care product is reformulated with natural colors and fragrances and with safer preservatives, it can affect the stability and viscosity of the entire formula – it is not just a simple one-to-one substitution. Formulators need to skillfully adjust each formula to account for these more natural ingredients, a process which is as much an art as it is a science. Wherever possible, Whole Foods Market continues to encourage manufacturers to formulate with the abundance of natural ingredients that are becoming more readily available. Moreover, as both manufacturers and consumers continue to request higher quality products and cleaner ingredients, the demand for these ingredients will increase, meaning they will become more readily available and more affordable, a benefit for all stakeholders. There are currently over 500 ingredients acceptable to Premium Body Care, with 400 unacceptable ingredients [1]; new ingredients are evaluated and added to the Premium Body Care ingredient list on an ongoing basis as the standard continues to expand and evolve.

Today, many new and existing manufacturers look to Premium Body Care as a formulation tool and guide – it is a highly utilized standard, looked upon as a model in the natural products industry. In many cases, Premium Body Care validated the existing internal standards for most vendors. However, because they came from Whole Foods Market, the standards have strengthened and broadened the credibility of the ingredient choices made. In terms of interacting with manufacturers, the company has emphasized to its vendors the importance of ensuring that label claims are clear, truthful, compliant, and not misleading. Whole Foods Market has also advised manufacturers to find trusted natural ingredient suppliers and to ask the right questions of their suppliers in order to get the right information for making formulation decisions. In guiding vendors to make higher quality and transparent personal care products, the company has highlighted the critical nature of going back through the supply chain to pinpoint where each of the individual raw materials comes from and how each is processed and manufactured. For this purpose, a "New Ingredient Form" has been developed to assist with the review of new ingredients. Throughout the process of culturing the program, Whole Foods Market has often taken the time to help smaller manufacturers, offering them basic technical guidance and referring them to outside experts as needed. Fortunately, a significant number of small, local vendors with beautiful products have been able to meet the Premium Body Care standards. The program has offered these companies not only an opportunity to gain recognition that puts them on an even footing with much larger names and product lines, but also one to grow their brand within the company.

By creating a positive incentive for moving to more natural, transparent and safer personal care products, the Premium Body Care program has been a success. As mentioned previously, Whole Foods Market did not want to prematurely require change; it was known that it could take many months to properly reformulate a product, and the company did not want any gaps on the shelves. Instead, Whole Foods Market wished to inspire and motivate change in partnership with their vendors. With time, Premium Body Care has caused a seismic shift in the supply chain within the company's vendor community, one that has been a huge win for customers and the environment. Currently, Premium Body Care is a trusted standard that has encouraged and accelerated the trend of reformulation to more natural products. It has been positively shown that it is possible to innovate and improve personal care product formulations by utilizing more natural and cleaner personal care ingredients; the goal moving forward is to refine this even further. That said, Premium Body Care is not a fixed standard, but a moving target, which will continue to evolve as new science-based studies and research come to light. The bottom line is that collectively many questionable ingredients have been eliminated from personal care products – both within and outside of Whole Foods Market – due to Premium Body Care.

12.3.3 WHOLE BODY RESPONSIBLE PACKAGING

Furthering Whole Foods Market's mission to support the sustainability of personal care products in their stores, in mid 2008 the company began to focus on product packaging. The company organized a series of meetings with its vendor partners to address the future of packaging for the body care departments and to develop guidelines for responsible alternatives to existing materials. Over the course of two years, guidelines were created in three primary material areas – glass, polymers (plastics), and paper – with the launch of the Whole Body Responsible Packaging Guidelines in 2009. In general, glass was given preference to plastic due to its recyclability, while paper with the highest recycled content or certified by an independent, third-party sustainable forestry organization was preferred. Plastic with the highest recycled content was also preferred. No new packaging with bisphenol A or with a GMO crop source biobased material was allowed after September 1, 2012. As with Premium Body Care, these guidelines were meant as a starting point for finding better alternatives, and as a moving target to be revised and improved accordingly, with the ultimate goal being zero waste. Despite being voluntary guidelines, much progress has been seen; many of the company's vendors have responded by creating products packaged in the most sustainable way possible.

12.3.4 ORGANIC BODY CARE LABELING STANDARDS

Whole Foods Market has worked to develop and implement Organic Body Care labeling standards. An ardent advocate for organic foods throughout its history, Whole Foods Market has helped to pioneer the organic movement, assisting in the development of the US standards that make Americans trust the meaning of the organic label. Company leaders have encouraged and supported more organic growers and manufacturers in the marketplace; they have also served as retailer representatives during the early development of shaping the organic standards and later to provide recommendations to the Secretary of Agriculture regarding organic production and processing on the National Organic Standards Board – a 15 member advisory board that guides the United States Department of Agriculture's (USDA) National Organic Program (NOP). Whole Foods Market continues to work closely with the USDA and the organic community to ensure that the organic standards remain strong and consistent with their shoppers' expectations. Applying this experience to ensuring that organic body care was properly labeled in its stores was a natural next step for the company.

With the belief that the definition of "organic" should not change substantially between the food and the body care aisles of their stores, the company introduced labeling guidelines for organic personal care products. Whole Foods Market implemented requirements that any personal care product making an organic claim – including products with "organic" in the brand name – should contain a substantial amount of organic agricultural ingredients, and should be certified by a third party. According to the USDA/NOP, in order for a food product to claim that it is "organic," it must contain at least 95% certified organic ingredients, with the remaining 5% tightly controlled. Unfortunately, there are no similar mandatory government standards for "organic" label claims on body care products in the United States; this means that consumers have little assurance that "organic" products contain any organic ingredients at all. This inspired Whole Foods Market to set its own set of standards for organic personal care.

The following requirements now apply to all cosmetics and personal care products sold at Whole Foods Market stores in North America:

I Any product that makes an "organic" product claim, like "organic shampoo" or "organic bath salts," must be 95% organic and certified to the same USDA Organic standard as for food.

II Products with more than 70% organic ingredients can make a claim such as "Made with organic essential oils and extracts" if they are USDA certified, or "Contains organic essential oils and extracts" or "Contains organic ingredients" if they are certified to the NSF/ANSI 305 Organic Personal Care products standard. The NSF/ANSI standard is a consensus-based private standard for organic personal care products that is similar to the "70% organic ingredients/Made with" level of the USDA standard, but it allows certain ingredients and processes that are specific to personal care products.

III Any products with "organic" in the product name must also be certified by a USDA accredited certifier or to the NSF standard.

Aside from these three certified claims, no other use of the word "organic" is allowed on the front panel of the packaging. In September 2012, the company announced that all organic personal care products in its US stores were third-party certified, making Whole Foods Market the first national retailer to mandate certification for all organic personal care products on its shelves. Plans to apply these standards to the Canadian stores are forthcoming. Whole Foods Market's efforts have been widely recognized [9], attesting to the standard's significance.

Implementation of these organic labeling standards was another huge step with a positive impact for all stakeholders. By requiring that organic products have substantial amounts of organic ingredients, the demand for such ingredients has increased, thereby increasing the amount of land farmed organically. Whole Foods Market has already seen many suppliers seek certification, increasing the organic ingredients in their products in order to comply, underscoring how this standard has been a huge win for organic farmers, the environment, and shoppers seeking more organic products.

12.3.5 INDUSTRY RECOGNITION

In December 2012, Whole Foods Market's "clear, forward-thinking policy" was nationally recognized in a retailer report published by the US non-profit Campaign for Safe Cosmetics – a national coalition of parents', women's health, environmental health, nursing, and consumer advocacy groups working to eliminate hazardous chemicals from personal care products. In the "Retailer Therapy" report [10], Whole Foods Market was named the highest ranked retailer for personal care product safety. The Campaign for Safe Cosmetics praised Whole Foods Market's proactive Premium Body Care and organic labeling guidelines, commending the company for implementing "a rigorous and transparent approach to screening the personal care products it sells." Confirming the company's solid commitment to promoting natural and organic personal care products, the Retailer report goes on to state "no other national retailer comes close to rivaling Whole Foods Market's leadership in the realm of safer personal care products."

12.3.6 CONCLUSION

It is important to emphasize the unique position of retailers as gatekeepers in the process of furthering the reach of natural personal care products. With Premium

Body Care, Whole Foods Market found a gentle way to encourage positive change among the manufacturers that provide products to its stores, which comprise a significant share of the natural personal care industry in the United States. Retailers can also play a critical role as educators. Whole Foods Market has invested resources in developing educational material for its team members, customers, and vendors; the breadth of its sphere of influence has allowed them to reach many people with information about Premium Body Care and about natural personal care in general. Finally, above all, Whole Foods Market has pushed forward with the Premium Body Care program with the purpose of acting as advocate for their customers. This is generally why Whole Foods Market is passionate about launching product standards that are clear and meaningful to their customers and setting the benchmarks for what they can expect within their stores. As evidenced by Premium Body Care, Whole Foods Market's key interest is to exceed their customers' expectations and make their stores the best and "safest" place to shop.

REFERENCES

[1] Whole Foods Market. Premium Body Care Unacceptable Ingredients List. Available from: http://www.wholefoodsmarket.com/sites/default/files/media/Global/PDFs/unacceptable_pbc.pdf (Last accessed December 20, 2012).

[2] Whole Foods Market. Store List. Available from: http://wholefoodsmarket.com/stores/list (Last accessed December 20, 2012).

[3] Whole Foods Market. Premium Body Care Website. Available from: http://www.wholefoodsmarket.com/department/article/premium-body-care-standards (Last accessed December 20, 2012).

[4] Whole Foods Market. *Be Good to Your Whole Body* Premium Body Care Pocket Guide. Available from: http://www.wholefoodsmarket.com/sites/default/files/media/Global/Departments/Department%20Article/Be%20good%20-%20BGTYWB/begood-0410.pdf (Last accessed December 20, 2012).

[5] Whole Foods Market. Premium Body Care Video. Available from: http://www.wholefoodsmarket.com/blog/whole-story/premium-body-care-quality-standards (Last accessed December 20, 2012).

[6] Villecco, J. (2009 Jan 29) Anatomy of a Shampoo. Whole Story Blog. Available from: http://www.wholefoodsmarket.com/blog/whole-story/anatomy-shampoo (Last accessed December 20, 2012).

[7] Villecco, J. (2010 Apr 23) Premium Body Care: The Vendor Connection. Whole Story Blog. Available from: http://www.wholefoodsmarket.com/blog/whole-story/premium-body-care-vendor-connection (Last accessed December 20, 2012).

[8] Villecco, J. (2012 June 19) How Much Do You Really Know about Sunscreen? Whole Story Blog. Available from: http://www.wholefoodsmarket.com/blog/whole-story/how-much-do-you-really-know-about-sunscreen (Last accessed December 21, 2012).

[9] Quenqua, D. (2010 July 14) Well, Is It Organic or Not? *New York Times*. Available from: http://www.nytimes.com/2010/07/15/fashion/15skin.html?pagewanted=all&_r=0 (Last accessed December 20, 2012).

[10] The Campaign for Safe Cosmetics (2012) Retailer Therapy: Ranking Retailers on their Commitment of Personal Care Product and Cosmetics Safety. Available from: http://safecosmetics.org/downloads/Retailer_Therapy.pdf (Last accessed December 20, 2012).

13

Targeting the Green Consumer

Kathy Sheehan

13.1 INTRODUCTION

Who is today's green consumer? (Figure 13.1) If an image of a tree-hugger wearing hemp clothing eating granola comes to mind, think again. Today's green consumer is more sophisticated and multi-faceted than these stereotypes would dictate. The green consumer also varies depending upon where in the world you are sitting.

And it is even too simplistic to say that green consumers in developed economies have similar attitudes compared to those in developing economies. Green concerns, behaviors, and attitudes tend to be quite localized as well. But despite these differences, there are a few key ideas to keep in mind. For all but the most green on a spectrum of green attitudes, green products must have the quality and efficacy that the consumer expects from a "non-green" alternative. The days of an altruistic consumer accepting an inferior product or experience simply because it is a green alternative are behind us. For many consumers, green is just as much a status symbol as it a core belief. And lastly, the easier you can make it for the consumer to go green without compromise, the greater the chances of success and breaking out of a "green niche" into the mainstream.

In a developed market such as the United States, some aspects of green have indeed gone mainstream. Since 1992, GfK Roper's Green Gauge study has surveyed Americans about their attitudes and beliefs about the environment and, in particular, how these beliefs translate into consumer behavior. In the past twenty

Sustainability: How the Cosmetics Industry is Greening up, First Edition. Edited by Amarjit Sahota.

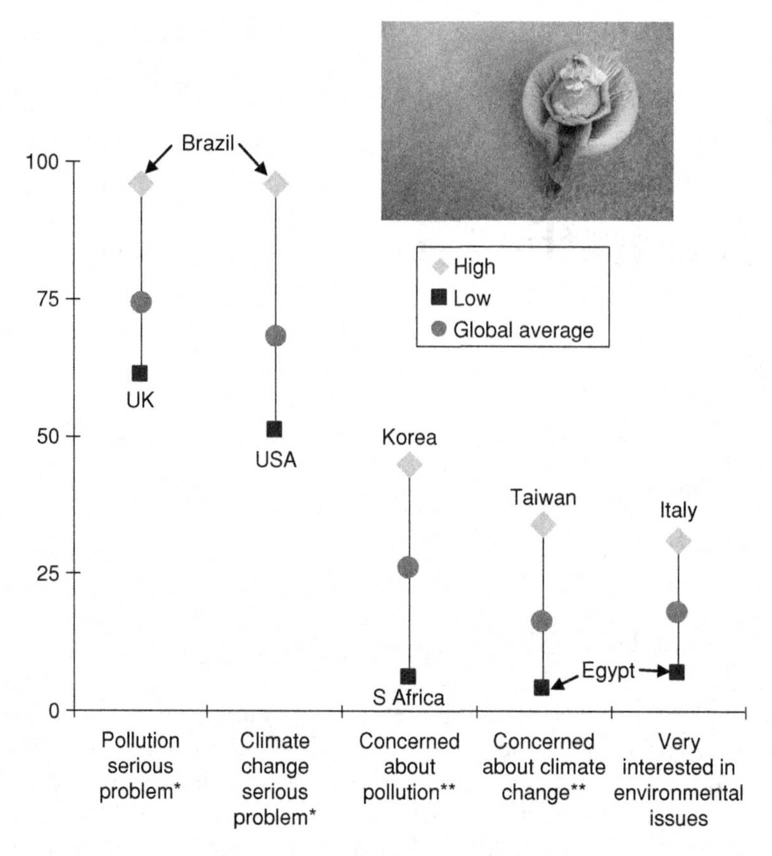

Figure 13.1 Who cares about the environment? Everyone, more or less. © GfK Roper Reports® Worldwide 2011.

years, we have seen environmental concerns ebb and flow, in large part based upon current events. For example, concerns about environmental pollutions spiked in 1989, clearly in reaction to the catastrophic Exxon Valdez oil spill, only to drop down again two years later. And, given the significant impact of the global financial crisis of 2008–2009 in the United States, many environmental concerns took a backseat to consumers' concerns about their personal finances and economic development (no help that many green products continue to be viewed as premium-priced by the majority of consumers). However, when one takes a longer view, we do see that certain behaviors have evolved, dramatically, since the 1990s. Twice as many Americans indicate that the recycle today than they did in 1990. Today, about 7 in 10 Americans feel that they know a lot or a fair amount about environmental issues and problems, up from 5 in 10 during the mid 1990s.

So, what is behind some of these development in the American market, and what does it offer lessons about the green consumer overall? The Green Gauge study

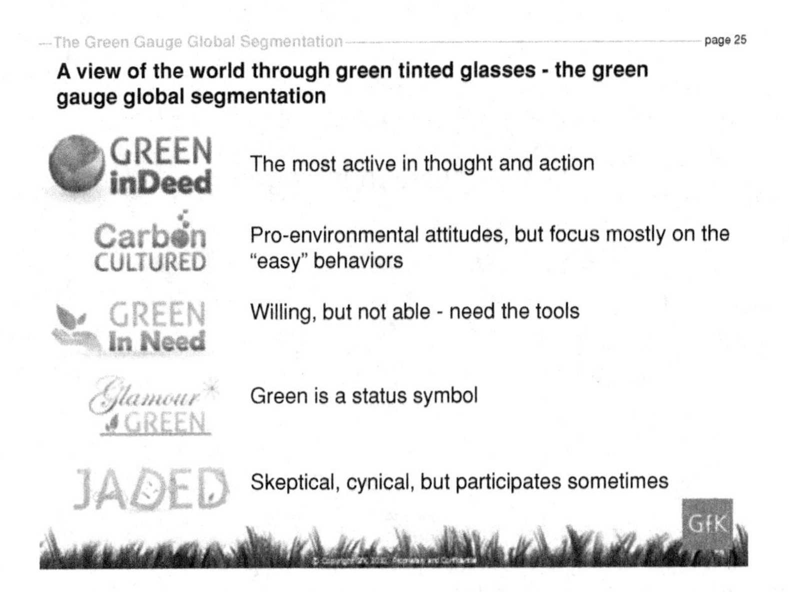

Figure 13.2 A view of the world through green tinted glasses – the green gauge global segmentation.
Source: © GfK Roper Reports® Worldwide 2011.

segments consumers in the United States and 25 markets around the world on a spectrum of "greenness." This segmentation factors in consumers' attitudes about the environment as well as their actual reported behaviors. Taking the entire population into account, the segmentation contains five, discrete consumer groupings (Figure 13.2).

The Green inDeed segment would score the highest on a scale of greenness. These consumers are the most active in terms of environmental attitudes, and are the most likely to carry these attitudes into the market place. The typical barriers to being green – knowledge about what to do and time to do it – are seemingly less of an issue for those who fall into this segment. This is a very active consumer segment – if they purchase a product that doesn't meet their expectations, they'll complain about it. And, because this is a group that is active with technology, they are more likely to reach a broader audience through word-of-mouth and social media.

Particularly prevalent in the developed markets of North America, Western Europe, and Developed Asia, where the environment has been on the national agenda for a while, the Carbon Cultured segment show a high degree of concern for the environment, but most of their behaviors still lag (Figure 13.3). They do things that are expected of them like recycling and conserving energy and water, but many of the "harder" behaviors are still at or below global norms.

In the developing markets of Latin America and Asia, we see a high prevalence of the Green In Need segment. Consumers in this segment have the desire, but lack

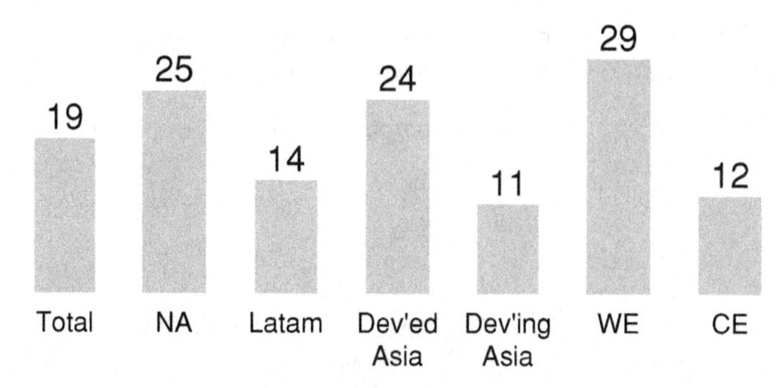

Figure 13.3 Segment snapshot: Carbon Cultured around the world.
Source: © GfK Roper Reports® Worldwide 2011.

the means and know-how, to be more green. Given the right tools, these people could climb the environmental ladder and become Green inDeeds.

Glamour Greens exist the world over (although they are more likely to turn up in urban areas). "Green" to this group is a status symbol. The environment is only skin deep to them. They do not have any core green values, but they will participate if it means they can display their green badge of honor.

Lastly, we have the Jaded segment. This is a cynical and skeptical group – they claim to be knowledgeable about the environment but are not concerned about it and do not want issues to stand in the way of economic progress. They have generally rejected the "green movement," and some may even be overtly hostile towards it. Still, they will participate if they have to (think mandatory recycling) or if doing so offers a tangible financial benefit (think saving energy).

Let's examine these segments within the United States as an illustrative example of who the green consumer is today and how that consumer is evolving. What do you need to know in order to tap into this mindset? What are some key considerations when it comes to the personal care and cosmetics category in particular?

13.2 UNITED STATES

In the United States, ease and availability of consumer products that play in the green space has been a hallmark of sustainability since the mid 2000s. From smaller brands such as BURT'S BEES®, to larger players such as Clorox's Green Works, green alternatives are increasingly available for the American consumer. (And, large acquisitions such as the 2007 Clorox acquisition of BURT'S BEES®, and Colgate–Palmolive's 2006 acquisition of Tom's of Maine, demonstrate one common way in which larger players have been venturing into the space.) Yet, at the same time that the US market has seen a virtual explosion in the variety and availability of green alternatives, we have seen some softening in Americans' enthusiasm about the green movement. Between 2008 and 2010, there was an 8 point *drop* in the number of Americans agreeing that global warming is a "serious issue" and between 2007 and 2010 a 15 point *increase* in the number of Americans agreeing with the statement, "First comes economic security, then we can worry about environmental problems." Clearly, the global recession, which was acutely felt in the United States, drives these changing attitudes. However, we also see evidence of "green fatigue" – increasing numbers of Americans reporting that environmentally friendly products are too expensive, are really not better for the environment, or cost too much (and sometimes all three). Several well-publicized product efficacy issues feed into this sentiment. For example, when the venerable Martha Stewart brand launched into green laundry detergent in 2010, product efficacy issues were widely publicized in traditional media and then picked up further momentum in social media as well. Consumer Reports' ranked it last in an evaluation of 50 laundry detergent brands, and stated it cleaned about the same as plain water.[1]

The USA is a somewhat polarized country when it comes to the environment, with about equal numbers of consumers falling into the "greener" Green inDeed and Carbon Cultured segments and the less green Jaded segment. Concern for the environment, while slowly growing, is well below the global average. It is perhaps not surprising that we see this dynamic when we think about how the green consumer has evolved in the United States. Two decades ago, the majority of Americans felt that the environment was a problem that needed to be tackled by large institutions – the government and business. There was little the individual could do. Today, that mindset has changed – consumers feel more empowered and recognize that they can be part of the solution (and the problem). Along with that empowerment comes an increasing sense of pragmatism and a desire to make sensible green choices. We also see that business is increasingly being viewed in a more favorable light when it comes to the environment (Figure 13.4). In 1990, 59% of Americans said a major reason for environmental problems is that companies do not develop and make available environmentally sound products, in 2011, that number went down to 45%.

On the one hand, we have an increasingly receptive audience for green products – recognizing that the individual *can* make a difference by their choices, and

[1]http://www.consumerreports.org/cro/video-hub/appliances/laundry–cleaning/martha-stewarts-dirty-laundry/16601904001/87908834001/.

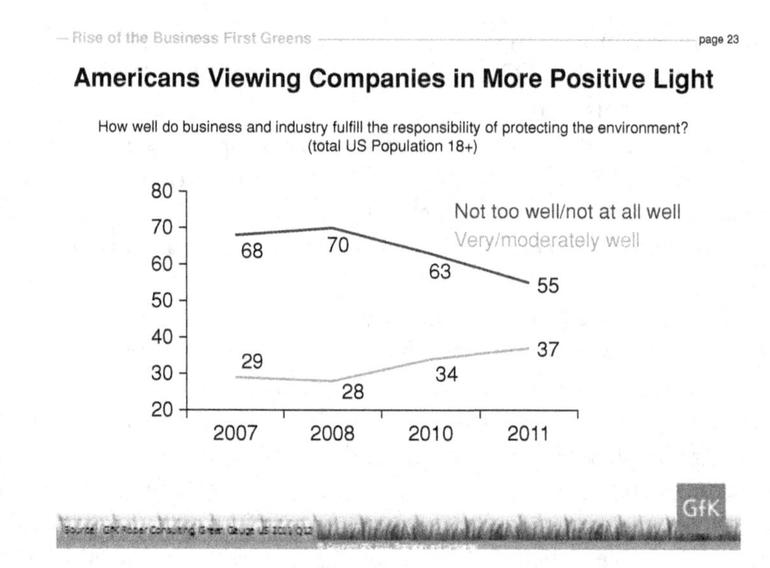

Figure 13.4 American's viewing business in a more favorable light.
Source: © GfK Roper Reports® Worldwide 2011.

feeling that business is helping them along in that quest. Yet, that same consumer is increasingly skeptical of the green offerings available in the marketplace and ever more critical in terms or product efficacy and price. Aside from quality, are there other levers that one should consider when thinking about the green consumer in the personal care category? The short answer to that is yes. When we survey Americans, who have purchased a green product in this category, about their motivators, the environment comes at the top. But personal health comes in at a virtual tie (Figure 13.5). So, while helping the environment may be the altruistic driver in purchasing green cosmetics, the health of the individual and their family, in many cases, is a primary driver. This learning supports the broader trends we see related to green product acceptance – that consumers in the United States are increasingly looking for "no compromise green" – products that are environmentally friendly, effective, priced reasonably, and offer other tangible benefits than simply being green. One marketer that had early success in this arena has been Method, the American-based personal care and home cleaning company. Method gained significant awareness in 2002 when its line was picked up in select Target retailers. The brand has successfully combined an environmentally friendly product, with attractive design and product packaging and fragrances. The idea was not only to make a product that was safer for the environment, but also to elevate home cleaning products – from something that was hidden under the sink to an item that the consumer would *want* to display on their kitchen countertop. In 2006, Method was ranked the seventh fastest growing private company in America by *Inc.* magazine, and has entered several international markets, including Japan and France.

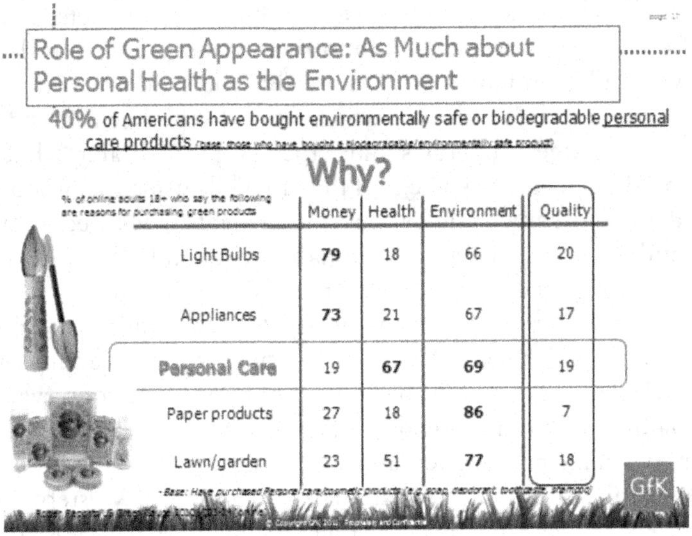

Role of Green Appearance: As Much about Personal Health as the Environment

40% of Americans have bought environmentally safe or biodegradable personal care products

Why?

% of online adults 18+ who say the following are reasons for purchasing green products	Money	Health	Environment	Quality
Light Bulbs	**79**	18	66	20
Appliances	**73**	21	67	17
Personal Care	19	**67**	**69**	19
Paper products	27	18	**86**	7
Lawn/garden	23	51	**77**	18

GfK

Figure 13.5 Motives to buy green products in the US.
Source: © GfK Roper Reports® Worldwide 2011.

Next we will turn to some other markets to see if these are current universals of the green consumer today.

13.3 WESTERN EUROPE

As a market, Western European consumers tend to lead on environmental attitudes and behaviors. One-quarter of Western European consumers fall into out "Green inDeed" segment, the highest proportion of any other region. A history of government regulation and the political climate no doubt drives this in many markets (as could be seen with the 2011 state elections in Germany, where the Green Party's vote tripled and Merkel's party lost power to a coalition including the Greens). And, in fact, one of the trends that we are seeing in this market is that consumers are calling for *more* government involvement. For example, 48% of Germans agree that "the government should be addressing our environmental problems, and not individual people like me" (up 4 points from 2010), which is a notably different mindset than the self-empowerment trend which we have discussed in the US market. However, empowerment is clearly evident when studying the Western European consumer, and we see overall concerns about the environment rising as well. For example, in Spain, 80% of the population reports that they conserve energy in the home (up 4 points from 2010) and in Sweden, 46% conserve water in their homes (again, up 7 points from 2010). Italy is one of the top markets in terms of conserving water, with close to 8 out of 10 consumers active in this arena.

Conservation is clearly a hot button issue in European markets, and can be used as a very effective way in which to tap into consumer's desires to be green. Germany's STW brand is an interesting case study. The product portfolio consists of shampoo, shower gel, and toothpaste which retails for between €8 and €15. Instead of a regular brand name, it operates under the "Stop The Water While Using Me!" message, reminding people to switch off the tap while using the products. All products are made with organic and/or natural ingredients, packaged in biodegradable containers, and manufactured with sustainability in mind. The company's web site reinforces the ease of the task – "Right there where it is used. Every day in the bathroom. At home, in restaurants and hotels. It's really quite easy to save water." Another campaign which started in the UK (and has now been imported into the USA) that is closely aligned with the European mindset of conservation has been Procter & Gamble's "Future Friendly." The web site offer tips to help consumers reduce their environmental impact (and save money as well).

Yet, despite this receptivity to sustainable products in Western Europe, there are still barriers in reaching this consumer. Similar to the USA, the developed economies of Western Europe were battered by the economic crisis of 2008, and continue to be impacted by government defaults. This, coupled with an aging population that will continue to place heavier burdens on social structures, makes economic growth a key concern with Europeans and one that is likely to grow in the foreseeable future. Europeans, along with Americans, index high on agreement with the statement that products are too expensive as a reason for not buying green. In this market, it is clear that players in the sustainability space must tackle the efficacy and price issues head on.

13.4 CHINA

One cannot have a discussion about cosmetics and sustainability without a thorough examination of China. As China continues its rapid rise, it is increasingly exerting its power and influence on the world stage in all areas – from politics, to economics, to the environment. (Indeed, in early 2012 *The Economist* magazine began to dedicate an entire section to coverage of China, the first such new section in 70 years – when they last created a section to focus on another market: the United States.) A key concern around the 2008 Beijing Summer Olympics was air quality – as the world descended upon China, it became abundantly clear that the environment was a concern to both those within and outside China's borders. Simultaneously, for many global marketers, China represents a land of immense opportunity. Understanding where sustainability is today and where it will be going tomorrow is key in this market.

It's definitely a focal area – the Chinese government's latest five-year-plan, released in March 2011, is considered by some to be the greenest five-year plan in China's history. Beijing adopted a lottery system that limits new issued car license plates to 240 000 in 2011, about one-third of the number registered in 2010. Beijing will apply a stricter emissions standard for the city's 5 million motor vehicles next year to reduce automobile exhaust, the biggest source of air pollution in the Chinese capital. China plans to cut its 2020 nuclear target after the post-earthquake 2011 crisis in Japan, and to build more solar farms. And the environment is on the minds of Chinese consumers as well. The largest proportion of Chinese consumers fall into the "Glamour Green" segment – green is a status symbol. One in four Chinese consumers fall into the Jaded segment. China lags in the more "green" Carbon Cultured and Green inDeed segments.

Although two-thirds of Chinese are concerned about pollution and climate change, they still lag in comparison to global figures. They also place more emphasis on economic security and well-being over environmental issues, and claim their busy lifestyles get in the way of doing more for the environment. Additionally, the Chinese feel they do not have adequate information to address environmental issues themselves.

Despite some negative attitudes, conserving water and energy are normal habits in China. And while recycling is still behind the global average, it became more common between 2010 and 2011.

Status is clearly a key driver in many categories in this market. (China is now the number two luxury market in the world and, by 2020, 44% of all luxury products globally will be sold there.[2]) And status is driving green purchases, particularly among the younger, more affluent urban consumers. For Glamour Greens, bold messages and products that let them display their environmental acumen will resonate the most. They want to go green with style. Think Eco-Fabulous.

NLGX is a small trendy clothing shop in downtown Beijing that offers recycled and sustainable products, including bags made from old newspapers. Owner, Ed Hung, says green is the new in-thing for younger urban Chinese: "Green is still quite new, in China, and even in the urban centers like Beijing. But I think from the success of these lines of bags that sell upwards of 1000 renminbi (USD 150) for a bag, people are starting to accept this and think of 'green' or 'huanbao' as more like a cool type thing. More like a cool image to portray."[3]

As in the United States, personal health is a key consideration with the purchase of many products. Chinese consumers tend to place very high importance on health and there is a strong affinity towards traditional approaches to health. Safety is a particular concern, given the environment as well as highly publicized product contamination incidents. Organic food, which is still niche in this market, has gained a lot of traction after multiple safety scares. Tofu Wang is one brand of organic soy milk whose brand positioning promises purity and no impact by pollution.

---

[2]Source: CLSA Asia Pacific Markets, As quoted in ChinaDaily.com, February 11, 2011.
[3]Source: www.nicelymadeinchina.com.

In terms of personal care in China, looking good is first and foremost about being healthy. China is one of the most health-oriented countries in the world – with more consumers here claiming to be focused on their long-term health. Chinese women see external beauty originating from inner well-being, and see healthy as beautiful. To them, improving the health of body, skin, and hair is much more fundamental to better appearance than external, temporary measures of using cosmetics or wearing nice clothes. No wonder the skin care segment still dominates the entire cosmetics and toiletry category in China. Color cosmetics or hair care products that promise to go beyond the basics and enhance the health of skin or hair are more likely to appeal (e.g., lipsticks enhanced with vitamins, foundations that promise to improve skin health). Inoherb skin care products are one such example that create the linkage: external beauty comes from the internal well-being.

Health lies in a combination of modern, scientific and traditional approaches. When it comes to health, Chinese women appreciate the modern, scientific approach. Yet, traditional Chinese medicine of hundreds of years is still held dear. Many use a combination of old and new. This merger approach applies to the health and beauty aids (HBA) category as well. There is a clear market for HBA products that combine traditional Chinese medicine with modern, scientific methods. Products with herb essence will also appeal to Chinese consumers' interest in natural ingredients.

Lastly, education is another lever (and potential barrier) in the green space in China. Fully 6 out of 10 Chinese consumers agree with the statement, "I would do more for the environment, but I don't know how," significantly higher than the global norm (51%). In addition, the fast-paced lifestyle of a rising economic super power is another barrier – agreement with the statement "I'm too busy to make my lifestyle as environmentally friendly as I would like" is on the rise among the Chinese.

As it is in so many other categories, China is a market with immense potential in the green cosmetics space. Linking health and safety with beauty is key to overcoming barriers that may be more challenging than in other developed markets. One such marketer who has recognized this is Jasmin Skincare. The Australian brand is marketing its organic skin care in China as "safe enough to eat." The products are certified organic and are marketed to Chinese consumers as safe beauty products, especially its Koala Baby Organics line.

13.5 LATIN AMERICA

Latin America is a market that is united by relatively high concerns about the environment, but not necessarily universally high actions. We also see a trend in individual empowerment and responsibility on the environment. For example, fewer Argentineans today agree with the statement "The government should be addressing our environmental problems, and not individual people like me" than even just one year ago. Although down significantly from 2010, two-thirds of consumers in Brazil agree that "I would do more for the environment, but I don't know how." As an emerging economy, Brazil comprises segments with pre-disposed barriers to a "green" mindset – nearly two in three are Green In Needs or Glamour Greens. The Glamour Green segment has grown in the past year as status has evolved more into the core mindset among Brazilians in the past decade, and "green" as a status symbol is now playing a larger role. Green skepticism is not widespread in Brazil; the Jaded segment is virtually non-existent.

Nearly all Brazilians show concern about pollution and global climate change – they in fact score highest among all markets on these two issues. They turn to both companies and the government to address the environment, as Brazilian consumers personally feel they do not have either the time or knowledge to address it themselves.

Water and energy conservation is prevalent in Brazil, but other green behavior, such as recycling, lags in comparison to global figures.

The "greenest" segment – the Green inDeeds – is the predominant mindset in Argentina, with about four in ten consumers falling in this segment. But similar to their Latin American counterparts, the Green In Need segment also shows high among Argentineans – suggesting that there are opportunities to expose them to even more green information.

Worries about pollution and global climate change are prevalent in Argentina, and Argentineans hold companies responsible for addressing environmental issues. But they also want to know how to address green issues themselves. Incorporating green in their purchase decisions and talking to others about the environment are more common behaviors in Argentina – the latter in particular can help pave the way for word-of-mouth opportunities.

Similar to other markets within Latin America, conserving water and energy are widespread practices in Argentina, while recycling is less common.

As we saw with China in the 2008 Beijing Olympics, the 2016 games in Rio de Janeiro will most likely vault Brazil (and by extension, Latin America) even more onto the global stage. It is perhaps not surprising, then, that one of the companies we are watching in the green space in Latin America has a strong global vision. MAOAR has the stated goal of becoming the leading Argentinean exporting company of organic, natural, and health products, and they actively look for distributors and wholesalers across the world that could make these products available to consumers. Their mission is to introduce Argentine organic products to the world as a solution to the global demand for quality and value. Continue to watch Latin America as a market with strong momentum in the green arena.

13.6 CONCLUSIONS

Today's green cosmetic consumer around the globe is perhaps just as varied as the consumer of any other product or service – influenced by unique motivations that may be personal, cultural, market-driven, or all three. However, there are a few commonalities that are particularly pertinent for the green cosmetics industry. As demonstrated in the United States, for better or for worse, environmental concerns tend to ebb and flow based upon current events. As the environment is now more and more in front of consumers – how can you effectively harness this increased awareness? Address the emerging price and efficacy issues head on – products must be competitively priced and work well. Access to green products is an issue in some markets as well. Greenwashing can seriously threaten the movement and destroy credibility. In our global village of 24/7 media coverage, social media, and instant access to information, transparency has become a commodity and there is nowhere for brands to hide. This is true for your brand, as well as the entire supply chain. Educate, educate, educate. Consumers continually cite lack of knowledge as a barrier to green living. Give them the practical knowledge they need to turn green thoughts into green actions. And lastly, don't forget about giving green some style. The Glamour Green segment dominates the global mindset and these consumers have at least some presence in every market. Give these consumers a chance to "show-off" their eco-consciousness – and help them broaden the movement.

14

Future Outlook

Amarjit Sahota

14.1 PREAMBLE

The preceding chapters of this book focused on specific areas of sustainability and marketing aspects. This final chapter summarises some of the key points covered in the book and gives future growth projections. Whereas many of the previous contributions provided best practices from industry, this chapter also highlights areas that warrant further investigation.

14.2 SUSTAINABILITY

Figure 14.1 gives the three pillars of sustainability, originally introduced in Chapter 1 (Section 1.2). In the last 10 years, sustainability has evolved from a concept into a discipline that has many facets. Although sustainability has become a multi-faceted discipline, most cosmetic and ingredient companies focus on its environmental aspects.

Large cosmetic companies usually take an integrated approach, whereby they instil elements of their sustainability plan into operational functions. For instance, Procter & Gamble plans to reduce by 20% its energy consumption, water consumption, disposed waste, as well as carbon emissions, by 2020. The implementation of this plan has involved changes to production processes, energy usage and waste

Sustainability: How the Cosmetics Industry is Greening up, First Edition. Edited by Amarjit Sahota.

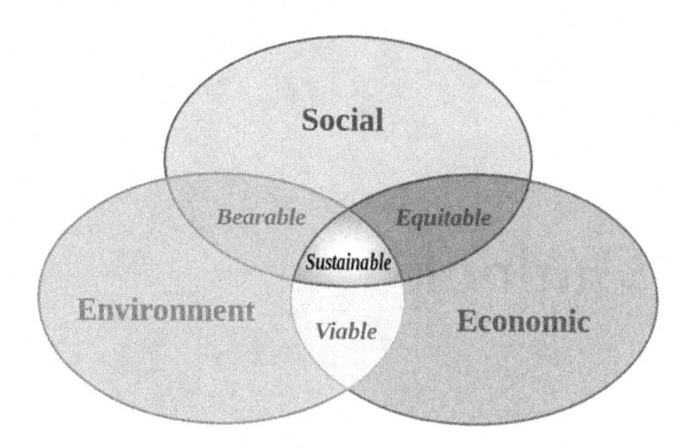

Figure 14.1 The three pillars of sustainability.
Source: Reproduced from http://en.wikipedia.org/wiki/File:Sustainable_development.svg, authored by Johann Dréo, inspired from image 'Sustainable Development'.

management. Social aspects of sustainability are not essential components of the sustainability plans of large cosmetic companies.

Small–medium sized companies tend to focus on specific areas of sustainability, such as ethical sourcing, carbon management or green formulations. The social aspect is also largely ignored unless the company is involved in ethical sourcing projects: for example, those involving fair trade practices.

Three reasons are believed to be responsible for the preoccupation with environmental impacts. First, consumers are well-versed in environmental issues, such as climate change, rainforest destruction and loss of biodiversity. Companies taking steps to tackle such issues are thus generally applauded by NGOs and consumers, even though the steps taken may be small. In comparison, social issues are more complex and not as easy to address.

Measurement is another factor: it is easier for a company to reduce its environmental footprint since this is relatively easy to measure and reduce. For example, a cosmetics company looking to become carbon neutral can undertake a life-cycle analysis (LCA) to measure its carbon emission. Steps can then be undertaken to reduce these emissions and offset the ones that remain. In comparison, the social dimension of sustainability involves elements that are not easy to measure or reduce. For instance, how do you measure the impact of corporate philanthropy: in terms of amount donated or how the donations are used?

Third, addressing social aspects requires 'soft skills' in organisations as opposed to hard (technical) skills. Soft skills or emotional intelligence are required to address social aspects like inequality in society/the organisation, human resources management, and working with NGOs/on community projects. Such emotional skills require nurturing and development, and cannot be learned at universities, whereas hard skills can. The requirements of these 'soft skills' make many companies shy away from addressing the social aspects of sustainability.

Lastly, many companies are not focusing on the social impact because it is not their sense of purpose. Many consider it is the role of government and charities to address social issues in the countries they operate in. Tackling environmental issues is a way of taking accountability for an organisation's actions, more so if they are involved in extracting and using finite raw materials and releasing waste into the environment.

Whereas few cosmetic and ingredient companies are focusing on the social aspect of sustainability, a major question mark hangs over the economic aspect: does it exist and, if so, how does it manifest itself? This book has given case studies and contributions that have centred on environmental and social footprints. The economic aspect has not deliberately been ignored; the industry has given little attention to this 'third pillar' of sustainability. Maybe it is because the general premise of enterprise/business is to make profits, so companies do not focus on this aspect? Some would argue that economic values are integrated into environmental and social aspects in terms of transparency, accountability and dealing with social inequality. In Section 14.8, the question will again be asked: what does the economic aspect mean in terms of sustainability?

14.3 SOCIAL DIMENSIONS

As stated in the previous section, the social dimension is not a focus when cosmetic companies undertake sustainability initiatives. If companies are to call themselves *truly* sustainable, they need to improve their social footprints. Figure 14.2 highlights elements of the social dimension that are pertinent to the cosmetics industry. Some pointers are given to companies looking at focusing on the social aspects of their business.

As stated in the previous section, a reason why cosmetic and ingredient firms do not focus on the social aspect is the difficulty in measuring and tracking social elements. For instance, how do you measure the results of 'fair' **human resources** policies: would they be measured in terms of gender equality, the number of employees from varying age groups, ethnic backgrounds, or disabled groups? It is common for companies to follow industry norms or government guidelines, rather than focus on such aspects in their sustainability programmes.

Rather then human resources policies, some companies are focusing on **employee remuneration** when looking at social equality. Whole Foods Market caps executive salaries at 19 times that of its average full-time workers. Dr. Bronner's Magic Soaps (another contributor to this book) caps executive salaries at 5 times that of its lowest paid workers. The Californian soap producer also puts 15% of employee

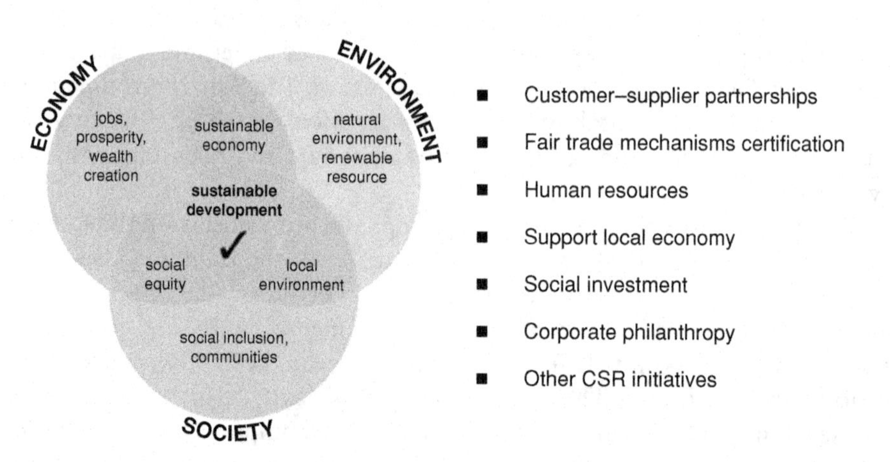

Figure 14.2 The social aspects of sustainability.
Source: Reproduced with permission from Organic Monitor © 2013.

salaries into a retirement plan, and up to 25% of salaries are paid as bonus into private health insurance [1].

The biggest social difference cosmetic and ingredient companies are making is arguably through **fair trade** sourcing projects. Indeed, many companies in the cosmetics industry have pioneered fair trade practices. Fair trade involves giving a fair price to growers, typically in developing countries, to protect them from the volatility of market prices. A premium is given to growers to help build social infrastructure, such as schools, hospitals, bridges, and so on. Many natural cosmetic ingredients are now available as certified fair trade; they include aloe vera, olive oil, coconut oil, shea butter, cocoa and honey [2].

The Body Shop set up one of the first major fair trade initiatives in the cosmetics industry in 1987. The Community Trade programme involved giving a fair price to a small group of Indian growers. Dr. Bronner's Magic Soaps has set up fair trade sourcing projects for organic olive oil from Palestine and organic coconut oil from Sri Lanka. Weleda has also set up similar projects with growers in Central Europe and Latin America, although the ingredients are not certified fair trade.

Cosmetic companies can also get involved in **social causes**. In 1997, The Body Shop launched a global campaign to raise self-esteem in women and to combat media stereotyping. The campaign focused on exceptionally skinny models because of the rising number of diet-illnesses such as bulimia and anorexia. The star of the campaign, a real-life size 16 plastic doll named Ruby, became internationally known. Ruby had a likeness to Barbie but with a voluptuous figure and luxuriant red hair; the campaign had the tag line 'There are 3 billion women who don't look like supermodels and only 8 who do'.

A number of cosmetic companies are involved in **corporate philanthropy**. The Avon Foundation for Women was set up by the American cosmetics company to champion social causes for women. Brazil-based Grupo Boticario has set up a foundation for biodiversity conservation projects. The Boticário Group Foundation protects over 11 000 hectares of Atlantic Rainforest and Cerrado, two of the most endangered biomes in Brazil. The foundation has donated over US $10 million in nature reserve programmes, enabling the discovery of 37 new species of plants and animals [3].

Avon and The Body Shop are considered two frontrunners in the cosmetics industry in terms of their social impact. Both companies have set up charitable foundations and have supported social causes for women. The Avon Foundation for Women tackles breast cancer and domestic violence. The Body Shop Foundation supports a number of environmental and social projects, including those involving human rights.

Weleda and Dr. Bronner's Magic Soaps are two natural cosmetic firms that have placed high emphasis on the social aspects of sustainability. Both companies have set up fair trade sourcing projects for their raw materials. Weleda has a positive and innovative human resources policy whereby its employees are rewarded by a profit-sharing scheme. Employees are also encouraged to undertake training courses for professional and personal development; for example, speech development and parental courses. Dr. Bronner's Magic Soaps has supported a number of environmental and social causes; its expenditure on such causes has roughly matched its total after-tax profits.

14.4 GREEN COSMETICS

Green cosmetics are referred to as natural and organic cosmetics: products that contain natural and organic plant-based ingredients and avoid synthetic chemicals, such as parabens, phthalates, sodium lauryl sulfate, and so on. The market for such products has increased from less then US $1 billion in the mid 1990s to US $9.1 billion in 2011 [4] (Figure 14.3). For over a decade, natural and organic was the fastest growing sector in the global cosmetics industry.

Although the major reason why consumers buy such products is health and safety concerns, sustainability has contributed to their success. In 2005, there was high consumer demand for natural and organic cosmetics; however, few products met consumer expectations as brands grappled with formulation issues. Natural and organic cosmetics failed to meet many consumer expectations because of performance and aesthetic issues. Many raw materials – especially alternatives to synthetic preservatives, emulsifiers and surfactants – simply did not exist. The

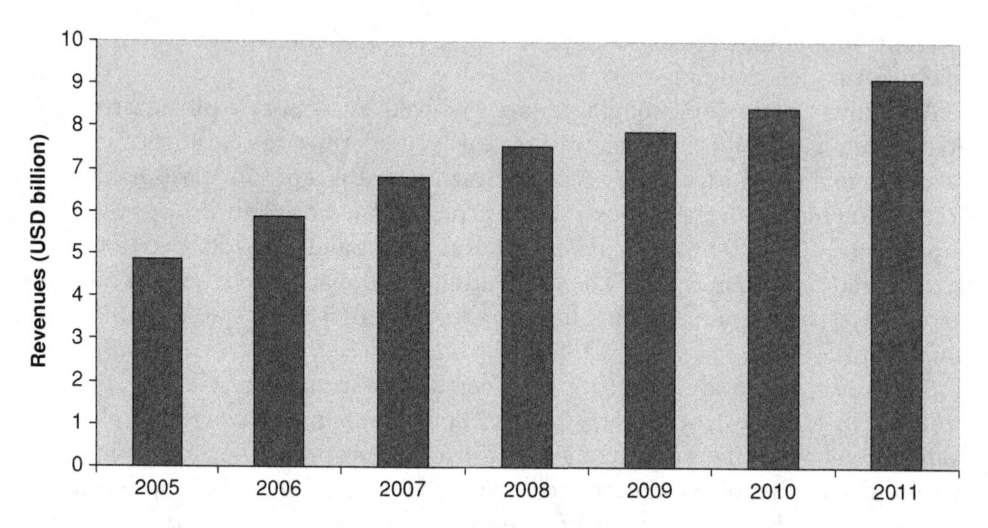

Figure 14.3 Growth of the natural & organic cosmetics market.
Source: Reproduced with permission from Organic Monitor © 2013.

alternative materials that were present could not meet the functionality of their synthetic counterparts.

As sustainability rose on the corporate agenda, cosmetic and ingredient companies invested in green formulations. A growing array of green ingredients eventually became available to formulators. This development encouraged innovation, expanding the range of natural and organic cosmetics in the marketplace. Furthermore, the quality and performance of products improved as more green ingredients became available. In 2012, all large speciality chemical companies had natural/organic cosmetic ingredients in their portfolios. A decade earlier, such companies were disinterested in this 'niche segment' with most raw materials supplied by small natural ingredient firms.

The availability of green raw materials from large chemical firms like BASF, Evonik and Rhodia encouraged large cosmetic firms to develop natural and organic cosmetic lines. Almost all such firms now have natural and organic lines. Johnson & Johnson introduced a dedicated range of natural personal care products for babies and children under the Johnson's Natural brand in 2010. In Europe, Beiersdorf launched the Nivea Pure & Natural range of natural skin care products in 2011. Unilever, Henkel and Garnier are also marketing natural/organic products under established brand names.

Consumer demand for natural and organic cosmetics has not been the sole reason why large cosmetic firms and chemical companies have taken the green route. As shown in Figure 14.4, such products have a much lower environmental footprint then synthetic products. By developing such products, companies can reduce their environmental impact.

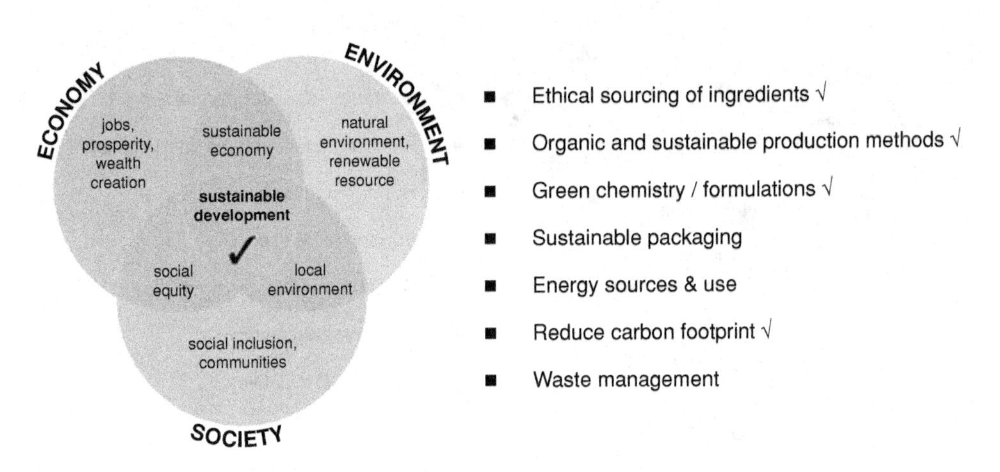

Figure 14.4 The environmental impact of fair trade cosmetics.
Source: Reproduced with permission from Organic Monitor © 2013.

It has become common for large cosmetic companies to give details of their green cosmetic ranges in their sustainability reports. It demonstrates how companies are adopting greener formulations and reducing their environmental impacts. An example is L'Oréal, which states it uses over 400 natural and organic ingredients in its cosmetic formulations.

Whilst natural and organic cosmetics enable companies to improve their environmental credentials, fair trade is also becoming prominent. Fair trade practices enable companies to improve their social footprints, as shown in Figure 14.5.

Growing consumer concerns about social issues in developing countries has been the major driver of the fair trade products market. The global market for certified fair trade products has grown from US $720 million in 2003 to US $6.4 billion in 2011 [5]. Although cosmetic products comprise a small percentage of total sales, the adoption rate of fair trade schemes is rising because it enables cosmetic companies to demonstrate their positive social impact to consumers. In the UK (the largest market for fair trade products), Boots, The Body Shop and Oriflame are some of the large cosmetic companies starting to brandish the Fairtrade mark on their products.

Many natural and organic cosmetic companies have adopted fair trade practices. Products with natural/organic symbols as well as fair trade logos are becoming more visible in the marketplace. Although this development is positive, since such cosmetic products have lower environmental and social footprints then conventional cosmetics, a concern is consumer confusion. Natural and organic cosmetic standards have a prohibited list of synthetic ingredients and processes to assure

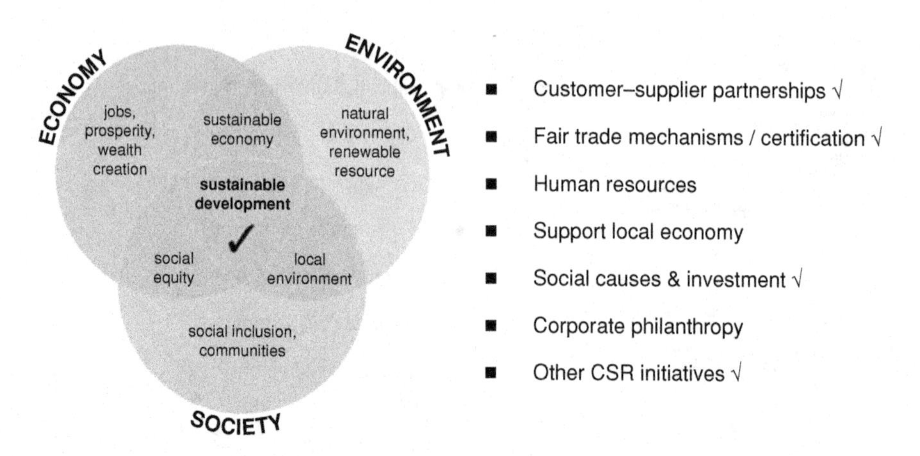

Figure 14.5 The social impact of fair trade cosmetics.
Source: Reproduced with permission from Organic Monitor © 2013.

consumers potentially harmful chemicals are not present in formulations. In comparison, cosmetic products can carry the Fairtrade mark if they contain a certain percentage of certified fair trade ingredients. There are no restrictions on the other ingredients in cosmetic formulations; thus, conventional cosmetic products with synthetic chemicals can carry the Fairtrade mark. Since consumers associate ecolabels with natural and organic cosmetics, they mistakenly think all products with the Fairtrade logo are natural/organic.

14.5 RESPONSIBLE CONSUMPTION

This book has shown how the cosmetics industry is becoming green by adopting sustainability practices. A number of case studies have been given on how cosmetic and ingredient companies are reducing the environmental and social impact of their products. However, little work has been done to encourage responsible consumption of cosmetic products.

A number of research studies show that the highest environmental footprint of certain cosmetic products is at the raw materials and consumption level. Figure 14.6 shows the results of one such life-cycle analysis (LCA) study by L'Oréal. Another study by the German company, Henkel, found that about 94% of the carbon footprint of its Schauma shampoo was at the consumption level. It found that 270 g of 290 g carbon emissions were produced during an average hair wash [6]. The biggest contributor is water usage, especially the process of heating water.

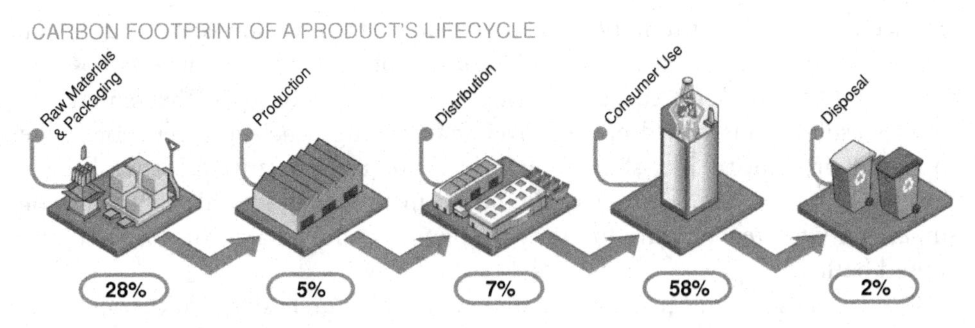

Figure 14.6 Carbon footprint of L'Oréal whole products portfolio.
Source: Reproduced with permission from © L'Oreal.

Small variations in water temperature and amount of water used had a significant impact on the carbon footprint of an average wash.

The challenge, therefore, is not only to produce more sustainable cosmetic products in terms of lower environmental and social footprints, but also to encourage responsible consumption.

Few cosmetic companies are encouraging responsible consumption of their products, possibly because such moves go against traditional 'marketing thinking'. Consumer goods companies have been intent on raising product sales, irrespective of how the products are used. For instance, retailers and brand owners have persuaded consumers to buy more products then they require by BOGOF (Buy One, Get One Free) and related promotions. It has also been to the advantage of hair care companies if consumers use excess shampoo during an average wash as it reduces the life-time of their products and shortens the sales cycle.

However, companies are slowly realising that they need to encourage consumers to be more responsible in their purchase, use and disposal of their products. Otherwise, no matter what sustainability initiatives are undertaken by companies, their products will continue to have high environmental footprints. The way forward is for cosmetic brands to work with their customers to take concerted efforts.

Unilever is one of the few cosmetic companies to address consumer behaviour in its Sustainability Living Plan. Introduced in November 2010, the sustainability plan set a number of targets to be achieved by 2020. Behaviour change is a key focus as Unilever has discovered that over two-thirds of the greenhouses gases and about half the water footprint of its products are at the consumption level. It has set a target of reaching 200 million consumers with products and tools that help them use less water while washing and showering by 2015. In November 2011, the company introduced its 'Five Levers for Change' which aims to encourage responsible consumption of its products. The starting point of the programme is to understand behaviour and what could motivate people to change [7].

Since responsible consumption has many facets, the question is: how far should cosmetic companies go? The previous example of shampooing suggests cosmetic companies could inform consumers on the amount of shampoo they should use when washing. This would prevent over-shampooing which requires more water to rinse hair, whilst under-shampooing would require more frequent washes and thus more energy use. However, should cosmetic companies also instruct consumers on what temperature to use when having a shower and for how long they should bathe?

There are also wider questions about responsible purchasing and consumption: should cosmetic companies encourage consumers to make less frequent visits to retailers to reduce their environmental footprint? Should they bring their own bags when purchasing their products? Should they encourage them to buy large pack sizes, even if they do not need them, so the packaging impact is lower? How should consumers reuse or dispose of their packaging?

Education is fundamental to changing consumer behaviour. If consumers realise they can make a difference by their purchasing decision, they are more likely to behave in a responsible manner. The exponential rise in fair trade product sales since 2003 demonstrates they are also willing to pay a premium for such beliefs.

14.6 ROLE OF GOVERNMENT AND LEGISLATION

One often asked question about sustainability is the role of government: what should the government do? Should the cosmetics industry be forced into sustainability actions by legislation?

The government and legislation play a very important role in encouraging positive change. For instance, the EU is implementing a ban on animal-tested cosmetics in March 2013. The ban is forcing other countries to look at alternatives to animal-testing methods, otherwise products made in those countries and tested on animals could not enter the lucrative European market.

Many in the natural and organic cosmetics industry call for EU legislation to protect the natural and organic terms. In the food industry, the organic term has been protected since the EU introduced standards for organic agriculture and organic foods in 1991. EU regulations ensure organic foods are produced according to a formal set of criteria. In June 2010, the EU logo for organic products was introduced. European legislation prevents fraudulent labelling and marketing of organic foods.

There are calls for similar legislation for natural and organic cosmetics because of the high level of marketing claims involving the natural and organic terms. It appears unlikely that such legislation will appear in the short term, considering the high number of private standards and differences between them. Furthermore, the adoption rates of such standards are relatively low. Organic Monitor research found that less than 2% of cosmetic products in Europe were certified according to these standards in 2011 [4].

The government can sometimes have a detrimental effect on industry development. The Brazilian government introduced biodiversity legislation to protect genetic assets of the country in 2001. Based on the Convention on Biological Diversity (CBD), the regulations restrict access to biodiversity resources and associated traditional knowledge and share the benefits resulting from its use.

The legal framework penalises Brazilian companies not obtaining approval when developing novel ingredients; however, it has not yet been enforced to foreign companies. Varying interpretations of the biodiversity legislation has created uncertainty, with companies like L'Oréal taking a cautious approach when sourcing raw materials from the country. The legislation was intended to protect biodiversity in Brazil, but has stifled innovation and created uncertainty [3].

Brazil is also one of the few countries to introduce a national regulation for organic cosmetics. Legislation prevents cosmetic companies making 'organic claims' on their products unless they meet national standards; however, the country has yet to introduce a standard for organic cosmetics. The absence of standards leads international organic cosmetic brands to market their products as 'natural' in the Brazilian market.

One area where the government can play an important role is encouraging companies to report on sustainability. A growing number of countries are forcing publicly listed companies in their countries to provide sustainability reports to improve transparency and corporate accountability. Such countries include the UK, France, Denmark, Sweden, Brazil and South Africa. Other countries, like Germany, have a voluntary code for sustainability reporting.

In summary, the government has an important role in encouraging sustainability. It is, however, important that legislation represents industry best practice and is not based on lofty, unachievable goals. The animal-testing methods example shows how legislation can be a force of positive change. Conversely, the Brazilian regulations for biodiversity and organic cosmetics show how the government can also play a detrimental role in industry development.

14.7 BENCHMARKING OF COSMETIC COMPANIES

This chapter has highlighted sustainability areas of improvement for cosmetic and ingredient firms. In spite of its historic association with unethical business practices, the cosmetics industry is doing fairly well in terms of sustainability performance. A couple of 2012 industry lists show how far cosmetic companies have come along the green road [8].

Seven cosmetic companies were listed in Ethisphere Institute's 2012 World's Most Ethical (WME) Companies list. According to the institute, 'the WME

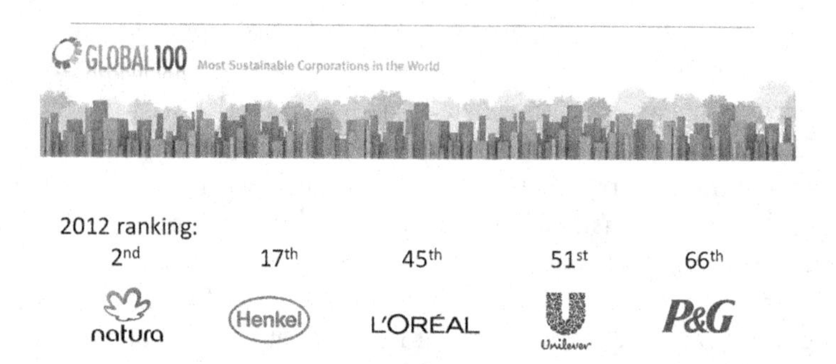

Figure 14.7 Cosmetic companies in the 2012 Sustainable Corporations list.
Source: Reproduced with permission from Corporate Knights and Organic Monitor.

companies designation recognises companies that truly go beyond making statements about doing business "ethically" and translate those words into action. WME honorees not only promote ethical business standards and practices internally, they exceed legal compliance minimums and shape future industry standards by introducing best practices today.' Cosmetic companies on the WME list were L'Oréal, Natura Brasil, Colgate–Palmolive, Henkel, Kao Corporation, Shiseido, and Kimberley-Clark.

Another assessment study by Corporate Knights named Natura Brasil as the second most sustainable corporation in the world in 2012. The Corporate Knights methodology involves evaluating corporations on a number of environmental, social and governance performance measures.

Corporate Knights praised Natura Brasil for its energy and waste productivity performance. The Brazilian cosmetics company has been carbon neutral since it introduced its carbon management plan in 2007. It is using the Greenhouse Gas (GHG) standard to calculate its carbon emissions, and then reducing and offsetting by investing in projects that have positive social/environmental impacts. Procter & Gamble, Henkel, L'Oréal and Unilever and were also listed in the top 100 sustainable corporations list (Figure 14.7). Number one in the 2012 sustainable corporations list was the Danish healthcare company, Novo Nordisk.

14.8 CONCLUSIONS

The cosmetics industry has traditionally received negative publicity for unethical business practices. Indeed, the industry still receives close scrutiny from NGOs and consumers because of the association of beauty with vanity. The previous section has shown that the industry is doing fairly well in terms of sustainability. Testament to this is that the second most sustainable corporation in the world is Natura Brasil (according to one industry assessment).

However, there is a long green journey ahead for the cosmetics industry. Many of the sustainability efforts have focused on environmental aspects, with little regard for the social and economic pillars. If companies are to call themselves *truly* sustainable, they have to take a holistic view and factor in the multiple facets of sustainability. Methods to improve the social impact of cosmetic products have been discussed in Chapter 3, as well as in Section 14.3.

Rarely discussed is the economic aspect, yet it is arguably the most important pillar of sustainability. In marketing classes, the price element is described as the most important 'P' in the 7Ps of marketing: price is the only element that generates revenues, whereas all other Ps incur costs. Similarly, the economic aspect of sustainability is considered the most vital. Without sound economic performance, companies will cease to exist if they do not deliver profits. In terms of sustainability, the fruits of the economic pillar enable companies to invest in environmental and social projects. Thus, the economic output of business provides resources to other pillars of sustainability.

Profit is not a word many environmentalists and green activists like to hear. However, profits are vital for sustainable development. Apart from allowing corporations to invest in environmental and social projects for sustainability, profits can have a direct impact on social communities. Profits enable businesses to sustain themselves, creating and maintaining employment. If profits are shared with employees and/or re-invested, they can also improve average incomes and standards of living. Thus, the economic pillar can indirectly bring positive social change. The large rise in GDP per capita in the Asian tiger economies in the 1980s and 1990s is testament to this. The average incomes of Singaporeans, South Koreans and Taiwanese citizens are now higher then those in many Western European countries. Furthermore, citizens in those countries are now as charitable and as environmentally conscious as those in the Western world.

Although economics provides resources, focusing on profits is clearly not the solution. The financial crises since 2008 have shown the damage that uncontrolled capitalism can create. In 2013, much of Europe remained mired in economic recession, whilst the USA was heading towards a financial cliff. Realising the limits of capitalism – especially social inequality and disregard for natural resources/the environment – there is a growing call for a new type of capitalism. At the World Economic Forum in January 2008, Bill Gates called for 'creative capitalism' whereby 'the power of the marketplace can help the poor'. Others are calling for a green economy: an economic development model based on sustainable development and knowledge of ecological economics.

The way forward appears to be capitalism with strong environmental and/or social values. This book has given case studies of companies/organisations that are leading in sustainability areas, such as ethical sourcing, sustainable packaging, energy and waste management, and so on. However, few companies are strong in many areas of sustainability aspects. Cosmetic companies that are 'all-rounders' in sustainability are those with strong ethical roots; they include Weleda, Aveda, BURT'S BEES® and Dr. Bronner's Magic Soaps. In certain cases, these companies exist to serve such ecological and social beliefs. For instance, Horst Rachelbacher founded Aveda in 1981 with the mission 'to care for the world we live in, from the products we make to the ways in which we give back to society'.

It is argued that large cosmetic companies have a greater impact in terms of sustainability because of their sheer size and influence on consumers. The likes of L'Oréal, Unilever and Procter & Gamble should indeed be commended for their efforts to reduce their sizeable environmental impacts. However, if we are to meet the challenges of the twenty-first century then a new way of thinking is required. Sustainability needs to become an integral part of these companies' business, rather than a support or auxiliary function. Companies like Aveda and BURT'S BEES® are leading in sustainability because it is part of their corporate DNA. Furthermore, such companies demonstrate that 'going green' or sustainability does not have to be at the expense of profitability. If anything, being green has been the cornerstone of their success.

This book has showcased leading companies and highlighted industry best practice in terms of sustainability in the cosmetics industry. It has been shown that sustainability is no longer a fad; it is being embraced by cosmetic and ingredient companies of all sizes. Although companies are taking different approaches to sustainability, the objectives are the same in that they want to reduce their environmental and social impacts. It is hoped that an outcome of this book is that the industry will be inspired towards positive change. The cosmetics industry maybe turning green, however there are many more shades to go!

REFERENCES

[1] Organic Monitor (2010), Strategic Insights report on CSR & Sustainability in the Cosmetics Industry, London, UK.

[2] Organic Monitor (2008), Strategic Insights report on The Potential of Fair Trade Cosmetics and Ingredients, London, UK.

[3] Sustainable Cosmetics Summit Latin America, Sao Paulo, September 24–26 2012.

[4] Organic Monitor (2011), The Global Market for Natural & Organic Cosmetics, London, UK.

[5] Fair Trade International http://www.fairtrade.net/.

[6] Case Study Shampoo by Henkel AG & Co. Kg AA, Pilot Project Deutschland, 2008.

[7] Unilever (2011), Unilever Sustainable Living Plan, Progress Report.

[8] Organic Monitor, London, UK. The Greening of the Cosmetics & Personal Care Industry Gathers Pace http://www.organicmonitor.com/r1104.htm.

Index

Sustainability: How the Cosmetics Industry is Greening up, First Edition. Edited by Amarjit Sahota.

Printed in the United States
By Bookmasters